Spacetime

Einstein's biggest Mistake

Amrit Srečko Šorli

Title: *Spacetime. Einstein's biggest Mistake*

Author: *Amrit Srečko Šorli*

ISBN: 978-1-63902-415-5

Cover image: www.pixabay.com

Publisher: Generis Publishing

Online orders: www.generis-publishing.com

Contact email: info@generis-publishing.com

To discover the unknown,

you need to step out of the known.

Amrit Srečko Šorli

Bijective Physics Institute

Slovenia

TABLE OF CONTENTS

Einstein has misunderstood time and space *1*

The time-invariant space model surpasses presentism and externalism *3*

In time-invariant space there is no time-travels and there is no time-symmetry . *9*

Entropy, gravity and entanglement in time-invariant space *11*

Timeless Superfluid Quantum Space as the Unified Field Theory *13*

SQS as the unified field model *14*

Complex time-invariant $\mathbb{C}^4$-SQS is the medium of quantum entanglement *20*

Advances of Relativity Theory *21*

Replacement of curvature of space with variable energy density of SQS *22*

The Lorentz factor and variable energy density of the SQS *24*

Advance of Special Relativity *26*

Bijective Advance of General Relativity *28*

Gravitational redshift *31*

Shapiro gravitational delay *33*

Gravitational lens *35*

Gravitational waves are waves of superfluid quantum space *36*

Black holes are rejuvenating Systems of the Universe *37*

Calculation of the "Schwarzschild energy density" *38*

Variable energy density of SQS at the distance d from the centre of the stellar objects *39*

The quantum mechanism of AGNs' jets *42*

Multiverse in Dynamic Equilibrium *43*

Cosmological redshift is "tired" light effect *45*

NASA's discovery means the end of expansion model of the universe and the end of inflation model *49*

Hubble law and Doppler effect in an expanding space *52*

Super-fluid quantum space, dark energy and dark matter *55*

CMB is the radiation of the existent universal space *57*

Big Bang cosmology and Einstein's steady-state cosmology have no answer about matter creation..*58*

Multiverse is in permanent dynamic equilibrium...*60*

Big Bang cosmology model timeline seems wrong...*61*

Bio-cosmology - Multiverse, Life and Consciousness......................................*63*

Quantum mechanics of life negentropy...*64*

Materials and methods..*66*

Discussion on obtained results...*70*

Evolution of life, order, disorder and randomness...*72*

Particle physics needs rigorous re-examination...*74*

Vortex model of elementary particles and the double-slit experiment.............*77*

The existence of antimatter in physical reality is questionable........................*78*

The Higgs mechanism is an unnecessary complication.....................................*79*

References..*82*

Einstein has misunderstood time and space

Back in 1905 Einstein arranged the marriage between time and space. This was the most unfortunate marriage ever in physics. It has caused the stagnation of physics for more than a century. Time has become part of space, they merged into spacetime. Gravity was understood as there result of spacetime curvature. NASA has measured back in 2014 that universal space has a Euclidean shape. Since then, the model of curvature of spacetime makes not much sense. Einstein has introduced the idea, that universal space (he named it "spacetime" is "empty", deprived of physical properties, it has only the geometrical structure – curvature which carries gravity. Reading this book, you will discover that the spacetime model was Einstein's biggest mistake. Still today mainstream physics believes that spacetime is the fundamental arena of the universe despite there is no single data confirming this convenience.

In this book, we will make research together and we will replace the model of spacetime as the fundamental arena of the universe is with the model of the time-invariant universal space, where time is merely the duration of change, i.e., motion. In time-invariant universal space time is not its 4th dimension. Experimental physics confirms this view, with clocks we measure the duration of a material change, i.e., motion in space. Moreover, the rigorous analysis of Special Relativity formalism of the fourth coordinate of spacetime confirms that the fourth coordinate X4 is not time t:

$$X_4 = ict,$$

In equation above i is the imaginary unit, c is the light speed, and t is the duration of photon motion in space. Time t is not X_4:

$$t \neq X_4.$$

Einstein has interpreted the time t as the 4th coordinate X_4 of a Minkowski manifold. He wrote: "If we replace x, y, z, $\sqrt{-1}\,ct$ by x_1, x_2, x_3, x_4, we also obtain the result that $ds^2 = dx_1{}^2 + dx_2{}^2 + dx_3{}^2 + dx_4{}^2$ is independent of the choice of the body of reference. We call the magnitude ds the "distance" apart of two events or four-dimensional points. Thus, if we choose as time variable the imaginary variable $\sqrt{-1}\,ct$ instead of the real quantity t, we can regard the continuum spacetime, in accordance with the special theory of relativity, as an "Euclidean" four-dimensional continuum, a result following by the consideration of the preceding section". In the above citation, Einstein suggests that we can choose the time variable t as the imaginary variable, can be written as follows:

$$t = \sqrt{-1}\, ct\,.$$

Equation above is false because on the left side of the equation we have t and on the left side we have $\sqrt{-1}\, ct$. Combining equation above with equation well known equation $X_4 = ict$ we get:

$$X_4 = itc^2\sqrt{-1}.$$

This equation is false and confirms that Einstein did a mistake keeping and interpreting time as the dimension of a four-dimensional continuum. Physics is still today suffering this misinterpretation of time that is solved in this article: time is the duration of material change, i.e., motion in time-invariant space.

Considering time as the duration of a change running in space, the universal space results as being time-invariant; the duration of a given event running in the universal space does not change in any way the physical properties of space and is not a part of the space. NASA has measured in 2014 that the universe space has Euclidean shape, measuring the angles between three stellar objects and getting 180° with 0.4% margin of error. This means that the universe has Euclidean shape and is infinite.

The idea of time-invariant structure of the universe was recently presented also by Hans J. Farr and Michael Hey: "One more cosmological possibility might perhaps need to be considered here, namely that the hierarchical structuring of masses in the universe which was considered in the above calculation could perhaps also be a time-invariant cosmic structuring, meaning that even though the universe undergoes an expansion in cosmic time, its hierarchical structuring endures or persists. Of course an expanding hierarchical universe must also change its mass density, however in such a way that the hierarchical structuring of matter persists, i.e. a time-invariant scale-invariance under these auspices must be considered". The right understanding of time as the duration of change, i.e., motion in time-invariant space is in our view one of the most important elements of 21st-century physics progress.

Today's cosmology examines the universe from the perspective of the universe is existing in some linear time that has physical existence. We are seeing the universe as something that has started long ago and is still developing in the present day. A rigorous examination of what is time confirms that time as the duration enters existence only when measured by the observer. There is no physical time running in the physical universe on its own. Universal changes are irreversible. When change X+1 enters existence, change X is not in existence anymore. When change X+2 enters existence, change X+1 is not in existence anymore. Changes run in a time-invariant universal space where there is no past, there is no present and there is no future. The linear time "past-present-future" exists only in the human brain, it has its physical origin in neuronal activity.

The only universe that exists is the one we can observe and measure. NASA has measured universal space has a Euclidean shape and is infinite. We are living in an infinite time-invariant universe where there is no physical past and there is no physical future. From this perspective, it makes no sense to build a hypothesis about the begging of the universe in some remote physical past because such a past is non-existent.

The universe is eternal and this eternity is NOW. Humans, we experience the time-invariant nature of the universal space as NOW. Eternity is not extending infinitely back into the past and is not extending infinitely ahead in the future. Past, present and future exist only in the human mind; eternity is NOW. Intuitively Einstein knew this and he expressed this in the following famous words: "People like us who believe in physics know that the distinction between past, present, and future is only a stubbornly persistent illusion". Still, in physics, he kept time as the fourth dimension of the model of the spacetime continuum. The spacetime continuum model has no correspondence in the physical world where we observe only material changes running in time-invariant space. Time as duration enters the existence when we me measure it.

The time-invariant space model surpasses presentism and externalism

Presentism believes that past and future are somehow coexisting in the present moment that is the only moment that exists. The entire history and the entire future of the universe are squished in the present moment. Eternalism believes that time is extending infinitely far into the past and infinitely far into the future. We can imagine presentism as the mathematical point and eternalism as the infinite straight line.

We will examine both views from the pragmatic view. You take a stone, you keep it above your leg, and then you throw it on your leg. When the stone hits you on the leg, you feel the pain. Before you lift the stone from the ground there was no pain in your leg. This proves that in the present moment events are not coexisting. They are following each order in the sequential order: 1. lifting the stone, 2. keeping the stone above the leg, 3. throwing the stone on the leg. We have seen in chapter 2 that sequential events in the universe are irreversible and are not coexisting. Presentism seems wrong; it is not the truth that everything coexists in the present moment. Eternalism sees the event with the stone is happening in the linear physical time one after other. Eternalism keeps the past as something real despite nobody ever reach into the past. For eternalism, the universe runs in some linear time that nobody ever measured and observed. Seems this view is not right.

In cosmology, presentism is an inspiration for the "block universe" model where everything that happens is coexisting. Eternalism is the inspiration for the "spacetime continuum" where we can have "closed time-lines" discovered by Kurt Gödel. The closed time-lines theoretically allow one could travel back in time and kill his grandfather and so it could not be born. That's why Gödel said: "In any universe described by the Theory of Relativity, time cannot exist". Gödel's discovery is still today interpreted wrong by some researchers who think that his development of General Relativity equations and consequently the discovery of closed time-lines indicates that one could travel back in time. On the contrary, Gödel was strictly against the idea of time travel. One can travel only in time-invariant space where there is no present, no past, and no future.

In the time-invariant space model, there is no physical time in the sense of present time as considered in presentism and there is no time in the sense of some linear physical time as considered in eternalism. Time as duration enters existence when measured by the observer. Time-invariant space that we humans experience as the present moment is the fundamental non-created eternal background of the universe. Universal changes run in this time-invariant space that is eternity itself. Humans experience the flow of changes in the frame of psychological time "past-present-future" and that's why we experience changes running in some linear physical time that is not there. We are "projecting" our psychological linear time that is the product of neuronal activity in the physical universe.

In the universe, we can only observe the relative rate of clocks and not some "relative time". Clocks run on the GPS satellites for 45 microseconds per day faster than on the Earth surface because of the General Relativity effect. And they run slower for 7 microseconds per day because of the SR effect. Clocks tick only in time-invariant space and not in some physical time. What is "relative" in the universe is not time, relative is the rate of clocks and velocity of material changes in general. A twin on the Moon would age faster than his brother on the Earth's surface because the velocity of changes on the Moon is a bit faster regarding the Earth's surface. The weaker is gravity faster is the rate of clocks and aging too. In interstellar space where gravity is weak, the twin would age a bit faster than his brother on the Moon surface or on Earth surface. But there is no "twin paradox". Twins are aging in time-invariant space.

In today's physics we still think that with rope we measure distance in space and with clocks we measure distance in time. Einstein has kept time as the dimension of a four-dimensional continuum. My research confirms that this spacetime continuum does not exist in the physical reality. Irreversible universal changes run in time-invariant universal space where black holes in the centre of galaxies are rejuvenating systems of the universe that is eternal and non-created.

Results of several researchers suggest that time has no physical existence, that it is an illusion. Bijective research methodology confirms their results are right: time is what we measure with clocks; we measure with clocks the numerical sequential order of material change, i.e., motion running in time-invariant space. Time as the duration of change enters existence only when measured by the observer. The change runs only in time-invariant universal space. Humans are experiencing a run of changes in time-invariant space in the frame of the linear psychological time "past-present-future" that has its basis in the neurological activity of the brain. Time-invariant space is the fundamental arena of the universe. In the universe, there is neither a physical past nor physical future. The universe is what we can observe with our senses and measure with apparatuses. All the rest is pure speculation.

Carlo Rovelli suggested that time is an illusion. On the other hand, he is not categorically denying the mainstream view that time cannot be eliminated from physics: "On the other hand, I also see well that the view I present here is far from being uncontroversial. Several authors maintain the idea that the notion of time is irreducible, and cannot be eliminated from fundamental physics". I will show in this book Rovelli is right about time being an illusion but still, time that we measure with clocks can remain in physics. The book introduces a model of time, which is an exact model of the time running in physical reality, using a *bijective research methodology* which is based on experimental results. There are no theoretical assumptions on time based on speculation, as for example what has been assumed for more than 100 years about time as the 4th dimension of space. There is no single experimental data confirming this last view.

Let's take the following example: a photon is moving in space from point A to point B. The distance d between A and B can be expressed as the sum of Planck lengths:

$$d = d_{P1} + d_{P2} + \cdots + d_{Pn} = \sum_{i=1}^{n} d_{Pi}.$$

Photon is moving from d_{P1} to d_{P2} and so on; every Planck distance d_{Pn} corresponds exactly to one Planck time t_P. In this sense, the Planck time is the fundamental unit of photon sequential motion from one to the next Planck distance. We can therefore write:

$$t = t_{P1} + t_{P2} + \cdots + t_{Pn} = \sum_{i=1}^{n} t_{Pi}.$$

$$\frac{\sum_{i=1}^{n} d_{Pi}}{\sum_{i=1}^{n} t_{Pi}} = c.$$

Photon is moving only in space and not in some physical time; the duration of photon motion from A to B in space is the sum of Planck times. Time is not continuous, is a discrete quantity; it is not some physical quantity that is running on its own, but the epiphenomenon of change, i.e., motion, its sequential numerical order. Time as the numerical sequential order of changes is a fundamental time; when fundamental time is measured by the observer, duration enters in existence. Duration is the so-called emergent time; there is no emergent time without the measurement of the fundamental time.

Back in 2005 I introduced a new research methodology, named *bijectivity principle*: let's consider the universe as a set X. In this set, we have four fundamental elements:

a) the energy in different forms: universal space is an energy structure, electromagnetic energy is a type of energy, dark energy is a type of energy, dark matter is a type of energy, energy in the form of matter;
b) the change;
c) the time;
d) the observer.

To build an adequate model of the universe in the set Y, which is the model of the universe, we must have there the same fundamental elements.

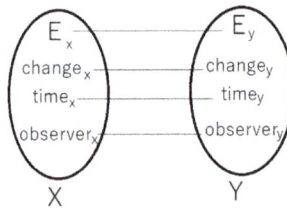

Bijective model of the universe.

$$X: \{E_x, C_x, t_x, O_x\},$$

$$Y: \{E_y, C_y, t_y, O_y\},$$

where E_x, C_x, t_x, O_x are energy, change, time and observer in the physical universe, and E_y, C_y, t_y, O_y the corresponding quantities in the model of the universe. We can write the following equation:

$$f(t_x) = t_y.$$

The time in the model of the universe t_y is related to the time t_x in the physical universe by the bijective function of set theory. Bijectivity principle assures an adequate model of the universe; the physical universe and the corresponding model of the universe are related by a bijective function. In this bijective model of the universe, time in the physical universe has exactly the same meaning in the model of the universe: time is the sequential numerical order of material change, i.e., motion. Time is not continuous, it is discrete; Planck time is the fundamental unit of time.

The idea that space and time should be defined on the basis of set theory is not new. It was already proposed back in 1997 by Costa, Bueno, and French: "Science is the search for structure. Such a bold claim certainly deserves further elaboration and justification, but it is one which, it has recently been emphasised, has been held by a number of scientists and philosophers. Our aim in this note is two-fold. First, we wish to relate the consideration of such an aim to a long-standing programme of work which has sought to develop a mathematically precise treatment of the notion of structure itself. In doing so, we shall analyse a particular illustrative case and thereby generate favourable evidence for the above claim as a whole. Thus, we shall consider the problem of constructing a set-theoretic structure, in Suppes's sense, which is capable of providing an axiomatic basis for those notions of space and time which underpin various theories of physics". We use in this research the bijectivity principle of set theory; the result is that universal space is a time-invariant. Time is the duration of material changes, i.e., motion in time-invariant space. This model fits experimental physics and is in agreement with the mathematical formalism of Special Relativity and with the neuroscience research results on time.

Time as some physical quantity that is running in the universe was never observed experimentally and represents a "hard problem" of physics. In order to understand time, and to build an adequate model of it, we have to understand how we experience change, i.e., motion.

The human being perceives with senses the information about the change. The information is transformed into an electromagnetic signal and moves through nerves to the sight center in the brain. Here there is a neuronal activity that is continuously creating the psychological sensation of linear time: past, present, future. Linear time "past-present-future" is the result of neuronal activity of the brain. Several researches confirm that animal and human experience of linear time has the origin in neuronal activity of the brain . We experience the material change, i.e., the motion in the frame of psychological time, which has its basis in neuronal activity; that is why we see changes running in some physical time despite there is no physical time in the universe.

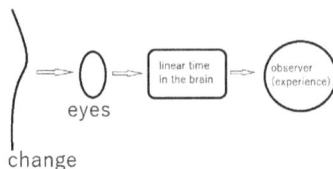

Linear time "past-present-future" is only in the brain.

Einstein was aware of the psychological time; he used to say: Past, present and future are stubbornly persistent illusions. Time has no independent existence apart from the order of events by which we measure it.". In Minkowski manifold, the arena of Einstein's Special Relativity, the 4th dimension is not time, but is a spatial coordinate $X_4 = ict$, i.e., the product of the imaginary number i, the speed of light c and the time t as the numerical sequential order of the change. Minkowski manifold has not dimension $3D + t$; it is $4D$. In this four-dimensional continuum, time is the sequential numerical order of events.

Recent research confirms that by zapping the areas of visual cortex V5 with magnetic field, the duration of a given event that people observe cannot be well estimated. The duration of a given motion is defined by the neuronal activity of the visual cortex. When functioning of the visual cortex area is disturbed by the magnetic field, the experience of motion is disturbed too.

In time-invariant space there is no time-travels and there is no time-symmetry

The time-invariant space is the underlying reality of the universe. In this non temporal underlying reality, time travels are categorically excluded. One can travel only in this underlying reality and time is the duration of its motion. In this perspective Minkowski manifold is only a theoretical model that describes this underlying time-invariant reality. The spacetime interval S has been defined as: $S^2 = c^2 t^2 - (X^2 + Y^2 + Z^2)$. With the introduction of natural units: $c = \hbar = 1$, the spacetime interval becomes: $S^2 = t^2 - (X^2 + Y^2 + Z^2)$ and the fourth coordinate of Minkowski manifold was interpreted as time t. The fourth coordinate of Minkowski manifold in its original form $X_4 = ict$ turned into $X_4 = t$. By taking natural units as 1, the time has been fully merged with space, and still today we think in physics that time is the 4th dimension of spacetime that is the fundamental arena of the universe. For a century it was thought this is one of the most important achievements of physics. Today this achievement is questionable. We have to admit that rigorous bijective mathematical analysis confirms that time t in Minkowski manifold represents the duration of photon motion in time-invariant space; not more and not less. The idea that time is the 4th physical dimension of the universal space is one of the main obstacles to physics progress. Universal space is time-invariant, there is no physical past and there is no physical future. Material change run in time-invariant space. This result is a challenge for cosmology. In the bijective model of the universe past and future are not coexisting in some eternal atemporal reality, they are non-existent. Irreversible universal change run in time-invariant space. Symmetry in time that is represented by the formula $T: t \rightarrow -t$ is the model that has no bijective correspondence with the physical world. In the universe, there is no symmetry in time because time is not a physical dimension in which the universe exists. Time is the duration of changes in time-invariant space of the eternal universe.

Time-invariant space is replacing spacetime as the fundamental arena of the universe. There is no such a thing in the universe as spacetime and "slices of spacetime".

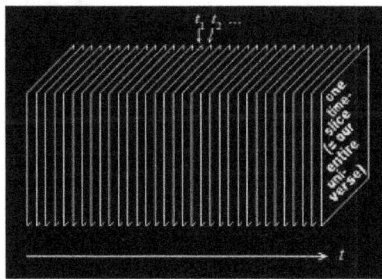

Hypothetical slices of spacetime

There is no such thing in the universe as some physical "time distance" Δt between two events:

$$\Delta t = t_2 - t_1,$$

where t_1 is a given slice of spacetime and t_2 is the next slice of spacetime. The duration between two events Δt enters into existence in the act of the measurement from the side of the observer. All events in the universe happen in the same identical time-invariant space. Clocks tick in time-invariant space. Clocks do not measure some physical time running on their own. Duration enters existence when measured by the observer.

time invariant space ⇨ change ⇦ measurement ⇨ time

Time is the result of measurement with clocks

The universe is time-invariant. It does not run in some physical time; it runs in time-invariant space. Universal changes are irreversible, physical past is non-existent. Future also is non-existent. Time travel is out of the question. One can only travel in time-invariant space.

Entropy, gravity and entanglement in time-invariant space

In physics, we experience the increase of entropy and run of physical changes through the linear psychological time of "past-present-future"; we experience that changes are running in some linear time. We have seen in chapter three that the idea that entropy of a given system is increasing in time is the wrong imagination based on our experience of material changes run in the frame of psychological time. Considering that universal space is time-invariant, the universal changes run only in this time-invariant space and not in time. With clocks, we measure duration of events in time-invariant space. The arrow of time exists only in the form of mathematical arrow of the sequential numerical order of universal changes running in a time-invariant space. The entropy of a given system is increasing only in space and not in time; time is the numerical sequential order of entropy increasing.

The idea that time has no physical existence and that with clocks we measure internal relations between different physical changes running in a timeless space is entering in the mainstream of physics, so as the idea that gravity does not require time and is encoded in a timeless configuration of the universal space. Time-invariant universal space is the direct information medium of entanglement by EPR-type phenomena. For a century, entanglement has been difficult to understand because of a wrong image, i.e., that spacetime (where time is a 4th physical dimension of space) is the fundamental arena of the universe.

It is shown in this book, universal changes are running in space that is time-invariant. Humans, we experience this time-invariance of universal space as "Now". Albert Einstein used to say: "that there is something essential about the Now which is just outside the realm of science". Neuroscience results that linear time "past-present-future" is created by the neuronal activity of the brain is bringing Einstein's Now into physics. Rovelli is right, time is an illusion in the sense, it has no physical existence. The mainstream did not understand well yet Rovelli's discovery. What is important to understand is that considering physical time is an illusion, clocks are remaining useful tools for measuring the duration of changes in time-invariant space. That the duration is the result of the measurement is a fact we cannot ignore. This fact will deeply change our notion of the universe, life, and human being.

The study of the universe in some linear physical time past-present-future seems not correct because we do not have single experimental evidence that physical time exists. Universal changes run in a time-invariant universal space. Time (duration) enters existence only when measured by the observer. Thinking time start running after the initial explosion seems not appropriate. Time does not run in the universe; it runs only in the human brain.

In this book, it is shown that as suggested by Barbour, Gomes, Fiscaletti, and Rovelli there is no physical time out there in the physical reality. Changes run in time-invariant space that is the fundamental arena of the universe. This seems an important result for the physics and cosmology progress.

"Time is duration" fits Newton's physics, Relativity, and quantum physics. Clocks without being seen by the observer are not measuring time, they are ticking in the timeless universal space. It is the observer's act of measurement that is creating time.

Timeless Superfluid Quantum Space as the Unified Field Theory

The novelty of 21st-century physics is the development of the "superfluid quantum vacuum" model, also named "superfluid quantum space" that is replacing spacetime as the fundamental arena of the universe. It also represents the model that has the potential of unifying four fundamental forces of the universe. Superfluid quantum space this book is represented as the time-invariant fundamental field of the universe where time is merely the duration of material changes.

Valeriy Sbitnev suggests that superfluid quantum vacuum also named superfluid quantum space is the physical origin of the universal space. Our research team has developed a model of the time-invariant n-dimensional complex superfluid quantum space which offers the new solution for Einstein's dream of a "Unified Field Model". In Einstein's Relativity the universal space is understood as a 4-D reality with tree spatial dimensions and one temporal dimension. Bezuglav also suggested that the superfluid quantum vacuum, which is the physical origin of the universal space, is four-dimensional. In experimental physics, time is duration of material change, i.e., motion in space. Taking this in account we developed the model of time-invariant n-dimensional complex superfluid quantum space, shortly "SQS".

The measured value of cosmological constant $\Lambda = 5.96 \cdot 10^{-27}$ kg/m³ is different from its calculated value following the Planck metrics for the magnitude of 10^{123}; this discrepancy is an unsolved subject of physics for decades. Regarding the suggested energy density of space proposed in this article, we are defending our proposal by the fact that the gravitational constant G is obtained by measurement and is expressed by the Planck energy density ρ_{EP} and the Planck time t_P as:

$$G = \frac{c^2}{\rho_{EP}\, t_P^2}.$$

This means that the Planck energy density ρ_{EP} reflects the real energy density of a 4-D universal space. In the absence of stellar objects, the energy density of the universal space has a value of Planck energy density which is $\rho_{EP} = 4.64 \cdot 10^{113} Jm^{-3}$. Meis has developed another formula for calculating the gravitational constant G:

$$G = \frac{l_P^2 c^2}{4 \pi e \xi},$$

where e is the elementary charge constant and ξ is the vector potential amplitude of the electromagnetic field to a single photon state ($\xi = 1.747 \cdot 10^{-25}$ V m^{-1} s^2). We can replace in equation above the term c^2 with the electric permittivity ε_0 and the magnetic permeability μ_0 obtaining:

$$G = \frac{l_P^2}{4 \pi e \xi \varepsilon_0 \mu_0}.$$

Equation above confirms that the 4-D SQS electromagnetic properties are defining the gravitational constant.

SQS as the unified field model

Time-invariant superfluid quantum space (SQS) has a general n-dimensional complex structure $\mathbb{C}^n$; every point of it has complex coordinates:

$$z_i = x_i + i \, y_i.$$

(x_i, y_i) $(i = 1, \dots, n)$ is an ordered n-tuple of real numbers $((x_i, y_i) \in \mathbb{R}^n)$; we consider its subset $\mathbb{C}^4$ where all elementary particles are different structures of $\mathbb{C}^4$-SQS and have four complex dimensions z_i. In $\mathbb{C}^n$-SQS the elapsed time of a given material change, i.e., motion is the sum of Planck times t_P:

$$t = t_{P1} + t_{P2} + \dots + t_{PN} = \sum_{i=1}^{N} t_{Pi}.$$

$\mathbb{C}^n$-SQS is time-invariant in the sense that time is not its fourth dimension. Material changes run in time-invariant $\mathbb{C}^n$-SQS and time is their duration. We do not have any experimental data that time is the fourth dimension of space and I suggest in this book a novel model where time is only the duration of change in time-invariant complex space. $\mathbb{C}^n$-SQS is the physical origin of the universal space. We call it "four-dimensional complex superfluid quantum space" ($\mathbb{C}^4$-SQS). Subatomic particles are different structures of $\mathbb{C}^4$-SQS; atoms, made out of subatomic particles, are three-dimensional physical objects, described by real geometry $\mathbb{R}^3$ and therefore follow the 3-D Euclidean geometry. Because of that we cannot fully grasp the complex subatomic level with 3-D apparatuses.

14

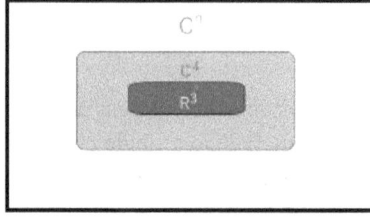

Structure of the $\mathbb{C}^n$-SQS universe.

The 4-D complex superfluid quantum space $\mathbb{C}^4$-SQS is the theoretical frame for the unification of gravity and the other three fundamental forces which have already been unified by the Standard Model. In this complex superfluid quantum space, we have four spatial coordinates which have a real and imaginary component. The energy density of $\mathbb{C}^4$-SQS is calculated in the terms of $\mathbb{R}^3$ matter in units kg/m³ and related to the mass m of a given physical object; every physical object with mass m is decreasing the energy density ρ_{Emin} of $\mathbb{C}^4$-SQS in its centre exactly for the amount of its energy:

$$E = mc^2 = (\rho_{EP} - \rho_{Emin})V,$$

where ρ_{EP} is the energy density of SQS far away of a stellar object in the interstellar space and V is its volume. By equation above we can calculate the minimal energy density of space in the centre of a given physical object:

$$\rho_{Emin} = \rho_{EP} - \frac{mc^2}{V}.$$

Equation above holds from the proton scale to black holes scale. Going away from the centre of a given physical object, the energy density of space is increasing by the following equation:

$$\rho_{Emin} = \rho_{EP} - \frac{3mc^2}{4\pi(r+d)^3},$$

where r is the radius of the physical object and d is the distance from its centre. When d tends to the infinite, ρ_{Emin} tends to ρ_{EP}.

15

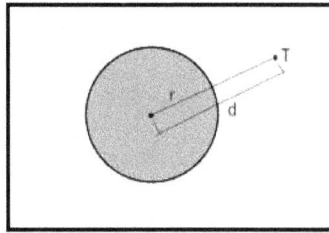

Energy density of $\mathbb{C}^4$-SQS in the point T at the distance d from its centre.

Gravity is carried by external pressure of $\mathbb{C}^4$-SQS towards the centre of physical objects; it is the force of $\mathbb{C}^4$-SQS pressure from the maximum energy density of $\mathbb{C}^4$-SQS towards its decreased energy density in the centre of the given physical object. Two physical objects are creating decreased area of $\mathbb{C}^4$-SQS energy density, causing outer pressure of $\mathbb{C}^4$-SQS towards its lower inner pressure. This outer pressure is gravity.

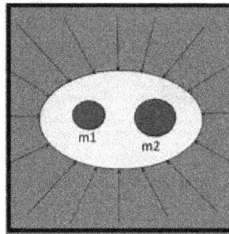

Gravity is the space pressure towards the physical objects.

We will calculate the energy density of space in the centre of different stellar objects, considering that these objects are non-rotating. In Table below there is the comparison of the energy densities of space in the centre of the black hole with the mass of the Sun, in the centre of the proton, in the centre of the Moon, Earth, and Sun:

Comparison values of the minimal energy density of space with respect to the centre of indicated objects.

Centre of objects	$\rho_P = 4.64 \cdot 10^{113} Jm^{-3}$
Black hole with mass of the Sun	$\rho_P - 1.58 \cdot 10^{36} Jm^{-3}$
Proton	$\rho_P - 5.43 \cdot 10^{34} Jm^{-3}$

Earth	$\rho_P - 4.94 \cdot 10^{20} Jm^{-3}$
Moon	$\rho_P - 3.00 \cdot 10^{20} Jm^{-3}$
Sun	$\rho_P - 1.26 \cdot 10^{20} Jm^{-3}$

The energy density of space in the proton centre is lower than in the centre of Sun, Earth and Moon because these stellar objects are made out of atoms where there is a vast empty space between the nucleus and electrons orbits. Proton's mass is very small compared with the mass of the Sun, but it diminishes the energy density of an extremely small area of space compared with that of Sun, that diminishes the energy density of an extremely big area of universal space; that's why the gravity force of the Sun has such a long-range.

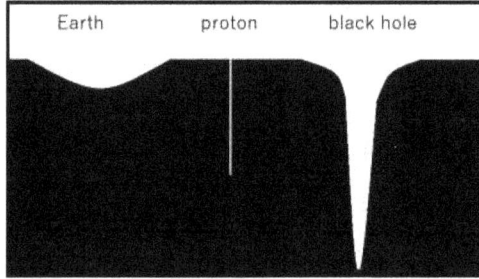

Energy density of $\mathbb{C}^4$-SQS in the centre of Earth, proton and black hole (figure is an approximation)

Proton has much lower energy density of $\mathbb{C}^4$-SQS in its centre than the Earth; however, it has almost no attraction force because of its extremely small mass. The calculation of attraction force because of lower energy density of $\mathbb{C}^4$-SQS in the centre of proton and neutron in deuterium nucleus is as follows:

$$F_g = \frac{1.67 \cdot 10^{-27} kg \cdot 1.67 \cdot 10^{-27} kg \cdot 6.67 \cdot 10^{-11} m^3 kg^{-1} s^{-2}}{(2 \cdot 0.87 \cdot 10^{-15} m)^2}$$
$$= 6.144 \cdot 10^{-35} N$$

where $1.67 \cdot 10^{-27}$ kg is the mass of proton and neutron, and $0.87 \cdot 10^{-15}$ m is their radius. This calculation confirms that gravity and strong nuclear force are not the same force as suggested by physicists Vayenas and Souentie.

We see that the energy density of C^4-SQS in the proton centre is higher for the rate of 10^2 than the correspondent energy density in the centre of the black hole. This shows that proton cannot be a mini black hole as suggested by Stephen Hawking. Voyager did not find these primordial black holes suggested by him.

From the macro to the microscale it holds that a given physical object is interacting with the C^4-SQS in which it exists and the result of this interaction are the inertial mass m_i and the gravitational mass m_g:

$$m_i = m_g = \frac{(\rho_{EP} - \rho_{Emin})V}{c^2}.$$

The rest mass m_0 of the proton is not its inertial mass m_i, but is related to the amount of C^4-SQS energy E which is incorporated in the proton:

$$m_0 = \frac{E}{c^2}.$$

The inertial mass m_i of the proton is the result of proton interaction with the C^4-SQS energy and the decrease in energy density of C^4-SQS in proton's centre is exactly for the amount of its mass and volume. This decreased energy density of C^4-SQS is the physical origin of proton's inertial mass. Einstein has proved that inertial mass and gravitational mass of a given physical object are equal. We have shown that they have the same origin in the decreased energy density of C^4-SQS.

In the model of C^4-SQS the electric field is the excitation of C^4-SQS on the three real dimensions X_1, X_2, X_3, and the magnetic field is the excitation of C^4-SQS on the tree real dimension X_2, X_3, X_4. Both fields have in common dimensions X_2 and X_3. The photon is then the excitation of C^4-SQS on X_1, X_2, X_3, X_4 dimensions, it is a 4-D wave of C^4-SQS; the light has a constant speed for all moving observers because it is a wave of C^4-SQS. When the source of light is moving closer to the observer or moving away from the observer the frequency of light will respect the Doppler effect. The source of light and the moving observer are all moving in the C^4-SQS. This model explains the physical meaning of the first postulate of Special Relativity, i.e., that the light has the same velocity for all observers; the light is a 4-D wave of C^4-SQS in which the observer and the 3-D source of light are moving. The motion of the observer or the motion of light source creates the Doppler effect but the light speed remains unchanged. In Special Relativity the photon is moving in a 4-D space of Minkowski, where time *t* is the element of the forth-dimension $X_4 = ict$. We have shown that time is just the numerical sequential order of material changes, i.e., motion in $\mathbb{C}^4$-SQS. When we measure the numerical order of photon motion from the point A to the point B in $\mathbb{C}^4$-SQS, we get duration. The photon is the wave of $\mathbb{C}^4$-SQS and does not move in some physical time.

The strong nuclear force is carried by gluons which bind together quarks inside the proton and neutron. Residual nuclear force between quarks is acting also outside protons and neutrons and hold them together in nucleus of an atom. In the model here presented gluons are excitations of $\mathbb{C}^4$-SQS and they represent 99% of proton mass. In this perspective, the proton mass can be seen as the excitation of the $\mathbb{C}^4$-SQS in the form of gluons and quarks. In Meis model the mass of electron m_e and mass of proton M_P are expressed by the physical properties of the electromagnetic vacuum:

$$m_e = 2\pi c e^2 \left| \frac{\xi}{\mu_B} \right| = 9.109 \cdot 10^{-31} \, kg$$

$$M_p = 2\pi c e^2 \left| \frac{\xi}{\mu_B} \right| = 1.672 \cdot 10^{-27} \, kg$$

The mass m_i of any elementary particle *I* can be expressed as:

$$m_i = 2\pi c e^2 \left| \frac{\xi}{\mu_i} \right|$$

With $|\mu_i| = \mu_B$ for the electron and $|\mu_i| = (2\alpha \,/\, n_i)$ for other particles, with n_i an integer and α the fine structure constant. In Meis model, as well in $\mathbb{C}^4$-SQS model, elementary particles are different energy structures of the $\mathbb{C}^4$-SQS energy. This is also the view of Erwin Schrödinger who used to say: "What we observe as material bodies and forces are nothing but shapes and variations in the structure of space". This is expressed in Einstein formula $E = mc^2$; E is the $\mathbb{C}^4$-SQS energy which is incorporated in a given physical object, *m* is the mass of the object.

Relativistic particles are interacting with the $\mathbb{C}^4$-SQS and additionally integrating $\mathbb{C}^4$-SQS energy into its structure. Relativistic energy *E* of a given accelerated particle is the sum of the rest energy E_0 and kinetic energy E_K which is incorporated energy of $\mathbb{C}^4$-SQS due to the motion of the particle:

$$E = E_0 + E_K = \gamma m_0 c^2 = (\rho_{EP} - \rho_{EminR})V,$$

where γ is Lorentz factor, m_0 is proton rest energy, ρ_P is Planck energy density, ρ_{EminR} is additionally diminished energy density of $\mathbb{C}^4$-SQS in the centre of the proton because the proton is additionally absorbing $\mathbb{C}^4$-SQS energy and so increasing its mass and energy, V is the volume of the proton at rest. The unification of electromagnetism and weak nuclear force into electroweak force was independently proposed by Sheldon Glashow, Abdus Salam and Steven Weinberg in the sixties of the last century; we introduced here a model where all four fundamental forces are carried by $\mathbb{C}^4$-SQS. Gravity force is carried by variable density of $\mathbb{C}^4$-SQS, strong nuclear force and electroweak force are carried by the excitation of $\mathbb{C}^4$-SQS.

Complex time-invariant $\mathbb{C}^4$-SQS is the medium of quantum entanglement

Several authors are proposing that entanglement is induced by gravity. On the other hand, there is a proposal that entanglement influences gravity. The result of our research is that gravity and entanglement are carried by the same medium which is SQS. In the model presented in this book gravity force between two physical objects does not induce entanglement and entanglement has no impact on gravity. Two entangled physical objects are entangled via SQS which variable energy density is also carrying gravity. Gravity does not influence entanglement and vice versa is also valid. In our model gravity and entanglement are both induced by the superfluid quantum space that is time-invariant.

The unified field theory of Albert Einstein is one of the main goals of modern physics. This goal can be achieved by the development of complex $\mathbb{C}^4$-SQS as the fundamental arena of the universe. Elementary particles and consequently strong nuclear force and electroweak force forces are different structures of $\mathbb{C}^4$-SQS. Gravity does not require the existence of some hypothetical particle graviton. It is carried directly by the variable energy density of time-invariant complex $\mathbb{C}^4$-SQS that is the medium of quantum entanglement EPR-type phenomena.

Advances of Relativity Theory

Advances of Relativity Theory are in the replacement of the spacetime model with time-invariant universal space that has a variable energy density. Every physical object with mass m and energy E is diminishing the energy density of space exactly for the amount of its energy. Lorentz factor has its origin in the variable density of universal space, we call it "superfluid quantum space" – SQS that is the primordial energy of the universe. Universal SQS is the absolute frame of reference for all observers as confirmed experimentally by the GPS system, which demonstrates that the relative rate of clocks is valid for all observers. A planet's perihelion precession and the Sagnac effect are the results of the SQS dragging effect. Development of Relativity Theory is based on three significant scientific discoveries:

a) spacetime is not a fundamental arena of the universe. Linear time "past-present-future" is phycological time based on the neuronal activity and exists only in the human brain. Irreversible universal changes run in time-invariant space. Time as the duration enters the existence when measured by the observer.

b) Entanglement happens in time-invariant space only and not in time. Time-invariant universal space is the immediate medium of quantum entanglement.

c) Universal space is not "empty", space is the fundamental energy of the universe, in today physics called "superfluid quantum vacuum" or "superfluid quantum space". We will call it in this book time-invariant superfluid quantum space – SQS.

GPS system proves that the relative rate of clocks on satellites relative to the Earth's surface is valid for all observers, including observers in aeroplanes, trains, ships, and cars. This experimental fact, along with everyday experience, suggests a revision to our understanding of the famous *Gedankenexperiment* of one observer at a train station and another observer on a passing train. Standard physics textbooks describe that a clock at the station runs faster for the observer on the train, and the clock on the train runs slower for the observer at the station. In classic relativity both observers have their own 'internal time' inside the reference system in which they exist and both have an 'external time' that exists in the other observer reference system. This interpretation features four distinct times: the proper time of the observer at the station, the proper time of the observer in the train, the external time of the observer at the station, and the external time of the observer on the train. On the other hand, GPS proves that the relative velocities of clocks at the station and on the train are equally related to the rate of clocks on orbiting satellites, so are valid for both observers. If this were not so, then GPS could not work properly. In this book, we will develop a model where the relative rate of clocks in all inertial systems depends only on the variable energy density of superfluid dynamic space (SQS) and is valid for all observers.

Replacement of curvature of space with variable energy density of SQS

Einstein tensor has three elements, curvature tensor on the left, Einstein constant and stress-energy tensor on the right side of the equation:

$$G_{\mu\nu} = \kappa T_{\mu\nu}.$$

Curvature tensor $G_{\mu\nu}$ describes curvature of spacetime due to the presence of a given mass that is expressed by the stress-energy tensor $T_{\mu\nu}$. Curvature tensor is useful only on the macro scale, it cannot be applied on microscale, for example proton. In this book curvature tensor will be replaced by the minimal energy density of SQS in the centre of the given physical object with the rest mass m_0. This formula is valid from the proton to the supermassive black holes (SMBH). Every physical object with energy E and mass m is diminishing energy density of SQS in its centre exactly for the amount of its energy and correspondent mass:

$$E = mc^2 = (\rho_{Emax} - \rho_{Emin}) \cdot V,$$

$$\text{in units: } J = kg \cdot \frac{m^2}{s^2} = (\frac{J}{m^3}) \cdot m^3$$

where ρ_{Emax} is density of SQS in interstellar space, ρ_{Emin} is density of SQS in the centre of a given physical object and V is the volume of physical object.

SQS model distinguishes between rest mass and inertial mass. A given physical object with the rest mass m_0 is diminishing the energy density of SQS in its centre exactly for the amount of its energy E. The diminished energy density of SQS is creating the SQS pressure in the direction towards the centre of the physical object. This pressure is the common origin of the inertial mass m_i and of the gravitational mass m_g of a given physical object. Einstein has proved inertial mass and gravitational mass are equal, we confirm in this book they are equal because they have the same origin.

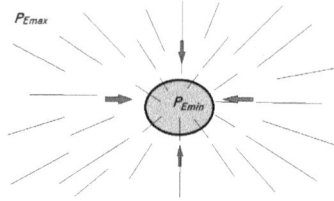

Inertial mass m_1 and gravitational mass m_g have the same origin in SQS pressure in the direction from ρ_{Emax} towards the direction ρ_{Emin}.

We can calculate the energy density of SQS at a given point on the distance R from the centre of a given stellar object as follows:

$$\rho_R = \rho_{max} - \frac{3\,m}{4\,\pi\,(r+R)^3} \, ,$$

where m is mass of the stellar object, r is radius of the stellar object and R is the distance from the centre of the stellar object to the point where we calculate ρ_R density of SQS.

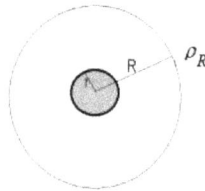

Energy density of SQS at the distance R from the centre

In Advanced Relativity curvature of space is replaced with variable energy density of space. More space is curved in GR, less is its energy density in Advanced Relativity. When R is zero, we have minimal energy density in the centre of a given stellar object and maximum curvature of space in GR. When R

is equal to the radius r of the stellar object, we have energy density of space on the surface of the stellar object. When R is close to infinity, we have the maximum energy density of superfluid quantum space (SQS) in interstellar space where the curvature of space in GR is at the minimum.

The Lorentz factor and variable energy density of the SQS

Lorentz factor γ expresses a diminished rate of clocks and a diminished velocity of material changes due to the motion. In the famous example of a train passing a station, t' is the elapsed time on the train and t is the elapsed time at the station, such that:

$$t' = \frac{1}{\sqrt{1-\frac{v^2}{c^2}}}\left(t - \frac{vx}{c^2}\right),$$

where $\frac{1}{\sqrt{1-\frac{v^2}{c^2}}}$ is the Lorentz factor γ, v is velocity of the train and x is the distance along the motion from the station clock to the clock in the train. This diminished rate of clocks on the train has its origin in the decreased energy density of the SQS inside the train. In general, a moving system interacts with the SQS energy so that the higher the velocity v, the stronger the interaction and more SQS energy is integrated into the moving object, which in turn increases its mass m of a moving object according to:

$$m = \gamma m_0 = m_0 + \frac{EK}{c^2},$$

where m_0 is the object's rest mass, EK is moving object kinetic energy in the form of integrated energy of SQS and γ is the Lorentz factor. The equation for the minimal energy density of SQS ρ_{Emin} in the rest wagon of the train is following:

$$\rho_{Emin} = \rho_{Emax} - \frac{m_0 c^2}{v}.$$

Formula for the energy density of SQS in the moving wagon $\rho_{Emin.m}$ is following:

$$\rho_{Emin.m} = \rho_{Emax} - \gamma \frac{m_0 c^2}{V},$$

where $\rho_{Emin.m}$ is the additionally diminished energy density in the centre of the wagon because moving wagon matter is absorbing some of the SQS energy which increases wagon's relativistic mass. This decreased energy density of the SQS $\rho_{Emin.m}$ causes the rate of the clock on the moving wagon to run slower. We can express the Lorentz factor as follows:

$$\gamma = \frac{(\rho_{Emax} - \rho_{Emin.m})V}{m_0 c^2}.$$

The difference in the energy density of SQS we can write as $(\rho_{Emax} - \rho_{Emin.m}) = \Delta\rho_E$. By replacing the $m_0 c^2$ with energy E of the rest object we get:

$$\gamma = \frac{\Delta\rho_E V}{E},$$

where E is the energy of the object at rest, V is the volume of the object and $\Delta\rho_E$ is the difference between the energy density of SQS far away from the physical object and the centre of the moving object. In equation above rest energy E and volume V of the object are not changing. The only parameter that changes the Lorentz factor is the diminished energy density of SQS in the centre of the moving object which depends on the velocity v of the object. So, the higher is the speed v, the stronger is the interaction of the object with the SQS, absorption of the SQS energy is greater, and the energy density of the SQS in the centre of the moving object becomes smaller. With a smaller density of the SQS in the centre of the wagon (and in any other moving object), the rate of the clock is slower:

increased velocity → increased absorption of the SQS energy → decreased energy density of the SQS → decreased rate of a clock.

In the SQS model, the relative rate of clocks and the relative velocity of material changes depend on the variable energy density of the SQS. In the SQS model, the relative rate of clocks and the relative velocity of material changes depend on the variable energy density of the SQS. For example, muon decay when approaching the Earth's surface decreases. The official explanation is that muons life-time is depending on the observer's reference frame. Official explanation says if you would be a potential observer moving along the muon, you would experience different muon decay as if you are on the Earth's surface. This seems wrong because the velocity of given physical phenomena has nothing to do with the observation. It depends only on the variable energy density of SQS. Coming closer to the Earth's surface muons enter the lower energy density of SQS, and their decay decreases. A muon's relativistic decay is valid for all observers and is determined only by the variable energy density of the SQS.

Advance of Special Relativity

In the areas of universal space where energy density of SQS is not changing the speed of light is constant for all observers because all observers exist in the same SQS and light is the vibration of the SQS. The velocity of light in the intergalactic space is constant, the energy density of SQS there is at the maximum. In the areas where the energy density of SQS is lower gravity is stronger and light speed diminishes minimally. We call this effect in classic relativity wrongly "gravitational time dilation"; what Shapiro has measured is that in stronger gravity light needs more time to travel on a given distance which means that its speed has a minimal diminishment.

The area of SQS around a given physical object is moving and rotating with it. We call this "dragging effect". SQS around the Earth is rotating and so the light motion needs a shorter duration when travels in the direction of Earth's motion because SQS is also rotating with the Earth. When light is moving in the opposite direction of Earth motion from B to A needs a longer duration. In both cases light speed is constant. By assuming constancy of light in stationary SQS and in moving SQS, we will develop an SR theory without contradictions as those that exist with the current SR in the thought experiment of two-photon clocks.

Here, we place two identical photon clocks on a moving train where one is positioned horizontally in the direction of motion, and the other is positioned vertically. According to the idea of "length contraction," the horizontal photon clock will shorten in length and tick faster compared to the vertically oriented clock that will not diminish in length. This scenario leads to a contradiction as SR does not predict that the two clocks in the same inertial system will have different rates. The solution is available through the development of an SR model with a Galilean transformation and Selleri's equation for the variable rate of clocks with no occurrence of "length contraction". Einstein's formalism of special relativity based on the standard Lorentz transformations may be derived from a more fundamental 3D Euclidean space, with Galilean transformations for the three spatial dimensions and Selleri's transformation for the rate of clocks:

$$t' = \sqrt{1 - \frac{v^2}{c^2}}\; t.$$

Selleri's equation confirms that rate of clocks is not related to the spatial dimensions.

A second contradiction occurs with the rate of the vertical photon clock on the moving train from the perspective of the observer at the station. The classical interpretation states that for the observer at the station, the vertical photon clock ticks slower because they see the photon in the clock moves in a 'zig-zag' direction.

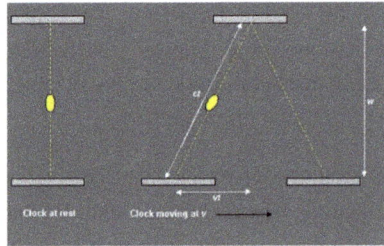

An observer at rest, seeing a moving clock photon.

This explanation may appear illogical because the optical illusion of the stationary observer cannot slow the rate of the moving clock. Instead, in the moving train, the energy density of the SQS diminishes, causing a reduced velocity of the photon. With the diminishing of SQS energy density also the velocity of light diminishes a bit. Therefore, the moving vertical photon clock ticks slower in the moving train because of the diminished energy density of SQS and not because of the optical illusion of the stationary observer in the station.

In Advanced Relativity rate of clocks in all moving inertial systems depends on the diminished energy density of SQS in the system due to its motion. The relative rate of clocks does not depend on the position of a given observer and is valid for all observers. GPS is proving this without any doubt. Because of the GR effect clocks on the satellites are running faster than clocks on the Earth surface for 45 microseconds per day. Because of the SR effect clocks are running slower on the satellites than the clocks on the Earth surface for 7 microseconds per day. This is valid for all observes.

If the clock would be taken out of the satellite it would keep the same rate. The mass of the satellite is too small to influence the rate of the clock because of the GR effect and SR effect. The only factor that determines diminished SQS energy density and so the Lorentz factor γ and consequently the rate of clock related to the SR effect is the velocity v of the clock.

In Advanced Relativity "length contraction" and "time dilation" where time is supposed to be the 4th dimension of space are abolished. We do not know a physical mechanism that would shorten the length of the objects that are moving in the direction of motion. The idea was created by Hendrik Lorentz in 1892 to save "ether". After Michelson-Morley's experiment has given a null result, Lorentz predicted that the beam in the interferometer that was pointed in the direction of Earth motion has shortened. In Advanced Relativity time is the duration of material change, i.e., motion in time-invariant SQS and cannot dilate. Time as duration is the result of the measurement form the side of the observer and as such has no physical existence. What is "relative" in the universe is the velocity of material changes that depends on the variable energy density of SQS.

Bijective Advance of General Relativity

In the model presented in this paper, the rotation of stellar objects also causes rotation of the surrounding SQS. For example, the rotation of the SQS around the Sun causes precession of the planets according to the following equation:

$$\sigma = \frac{24\pi^3 L^2}{Tc^2(1-e^2)},$$

where the perihelion shift σ is expressed in radians per revolution, L is the semi-major axis, T is the orbital period, c is the speed of light, and e is the orbital eccentricity. The mass of the Sun is not included as there is also no mass of a planet, so these masses do not affect the precession of the planets. In the model of SQS, the perihelion shift σ depends on the rotation of the SQS caused by the rotation of the Sun, which in turn pushing the planets and causes a perihelion

28

precession. With increasing distance from the Sun, the impact of the rotating SQS on planets diminishes along with the precession of the perihelion.

Irregular and spiral galaxies comprise approximately 60% of all galaxies in the universe. In the centre of most spiral galaxies exist a rotating black hole. I suggest that rotating black holes are rotating the surrounding SQS. This might be one of the physical causes of their spiral shape; dragging effects between the rotating black hole and rotating SQS diminishes with the distance from the black hole leading to the spiral geometry. The development of the mathematical model of this effect is one of the goals of our further research.

In 2019 NASA has reported: "as if black holes weren't mysterious enough, astronomers using NASA's Hubble SQS Telescope have found an unexpected thin disk of material furiously whirling around a supermassive black hole at the heart of the magnificent spiral galaxy NGC 3147, located 130 million light-years away. The conundrum is that the disk shouldn't be there, based on current astronomical theories". My proposal to solve this conundrum is that in current astronomical theories, supermassive black holes rotate in an "empty space." In SQS model presented in this article, supermassive black holes rotate in the medium of SQS, and their rotation, in turn, rotates the SQS. Therefore, this dragging effect of SQS might be a physical cause of thin disc that is furiously whirling around a supermassive black hole of the spiral galaxy NGC 3147.

In GPS, Sagnac effect corrections make the system work. Essentially, a signal when moving from A to B in the direction of Earth's rotation needs less time compared to when moving from B to A in the direction opposing Earth's rotation. In a SQS model, light has a constant speed regardless of the SQS's motion. So, when moving from A to B, light needs less duration (time) because it is moving in the same direction as the SQS.

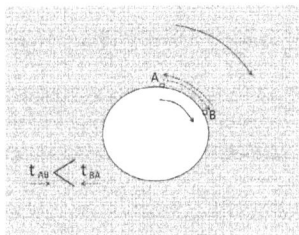

A light signal's duration due to the rotation of the quantum SQS.

Sagnac's experiment with the rotating interferometer is indisputable proof that photon does not move in the empty space deprived of physical properties. On the contrary, it proves that the photon is the excitation of the SQS that is dragged by the rotating interferometer.

The Michelson-Morley experiment demonstrated a null result because the area of the SQS around the Earth is not only rotating with the Earth but is also moving with the Earth.

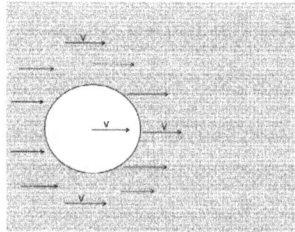

The SQS moves with the Earth.

So, the negative outcome of Michelson-Morley abolished the ether model. According to bijective research methodology, where every element in the model has exactly one correspondent element in physical reality, universal space is neither filled with ether, nor is it "empty". Instead, universal space contains material objects that contain energy. Energy and matter cannot exist in an 'empty' space deprived of all physical properties, so in this book universal space. We name it SQS, it is understood as the primordial energy of the universe. A photon is a wave of the SQS, and the velocity of this photon wave is the speed of light, c. The photon velocity c is invariant with respect to the SQS's motion as it appears through the Sagnac effect. The photon velocity diminishes minimally when a photon moves through a stronger gravity where the density of the SQS is lower, as is the case with the Shapiro experiment.

Motion and rotation of the universal space with physical objects is referred to as the "spacetime dragging effect". Dragging effect was measured by Josef Lense and Hans Thirring in 1918 and was called "frame-dragging" due to the belief that spacetime being distorted by rotating objects, reference. Recent research confirms that this "spacetime" model has no physical reality, so it cannot be dragged by rotating or moving objects. According to bijective research methodology, an adequate term would be the "SQS dragging effect".

30

In Advanced Relativity, the idea of "length contraction" and "time dilation" do not exist. Length contraction in SR is only a mathematical tool with no physical reality. By using "length contraction," Einstein achieved a constancy of light in all inertial systems. On the other hand, all observers measure the same value for the velocity of light because the light is the vibration of the SQS in which all observers move. Also, time, being in Special Relativity the fourth dimension of space, does not "dilate"; relative velocity of material changes (the rate of clocks included) depends on the variable density of the SQS.

Gravitational redshift

Gravity has its origin in the variable energy density of SQS. SQS fluctuations are directed from the higher energy density of SQS towards the lover density of SQS. Therese fluctuations that interact with photons to diminish their frequency, which is referred to as "gravitational redshift". When light from distant galaxies reaches the Earth, its frequency is lower. On its path to Earth, light loses some of its energy because it is moving against the SQS fluctuations that points toward the direction of galaxies, so that

$$E_{photon.Earth} = E_{photon.galaxy} - \Delta E,$$

where $E_{photon.galaxy}$ is the energy of the photon at the galaxy, $E_{photon.Earth}$ is the energy of the arrived photon at the Earth, and ΔE is the loss of energy due to the fluctuations of the SQS,

$$\Delta E = h\Delta v,$$

where h is Planck's constant and Δv is the decrease of the photon frequency due to SQS fluctuations.

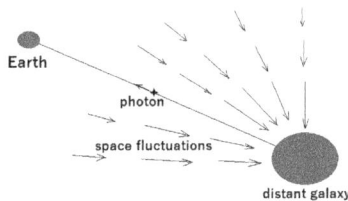

The redshift of light arriving from galaxies caused by SQS fluctuations.

Because of different densities of the SQS, the frequency of light also changes when moving from the source to the receiver above the Earth's surface. In a Harvard University experiment, a source on the Earth's surface and a receiver at the height of 22,5 meters were positioned.

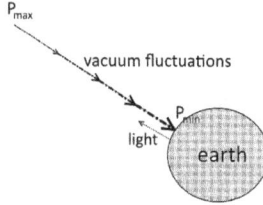

The redshift of light moving from the Earth's surface upwards.

The Mössbauer effect was used to measure the difference between y-ray emission and absorption frequencies at each end of the experiment. The measurement accuracy was $\Delta\omega/\omega \approx 10^{-15}$, which shows a change of light frequency as:

$$\frac{\Delta\omega}{\omega} = \frac{GM}{R^2c^2}h \, ,$$

where M and R are the mass and radius of the Earth, respectively. We can substitute into equation above for the Earth mass M with the $\frac{(\rho_{Emax}-\rho_{Emin})V}{c^2}$ and we get:

$$\frac{\Delta\omega}{\omega} = \frac{G(\rho_{Emax}-\rho_{Emin})V}{R^2c^4}h,$$

which can be expressed as:

$$\frac{\Delta\omega}{\omega} = \frac{G(\rho_{Emax}-\rho_{Emin})4\pi R^3}{3R^2c^4}h$$

$$\frac{\Delta\omega}{\omega} = \frac{4\pi RG(\rho_{Emax}-\rho_{Emin})}{3c^4}h.$$

Equation above confirms that gravitational redshift depends on the minimal energy density of the SQSρ_{Emin} in the Earth's centre. SQS fluctuation in the direction from ρ_{Emax} towards ρ_{Emin} are explaining so called "tired light" model of astronomer Fritz Zwicky. Zwicky proposed that light is losing some of the frequency when travelling vast distances from the galaxies to the planet Earth.

In model of Relativity here presented SQS fluctuations from ρ_{Emax} towards ρ_{Emin} are causing the Pioneer anomaly which means the observed deviation from predicted accelerations of the Pioneer 10 and Pioneer 11 spacecraft after they passed on their trajectories out of the Solar System. We suggested in out model that SQS fluctuations represent the barrier for the photons and also represent the barrier for the Pioneer spacecraft slowing down its acceleration.

Shapiro gravitational delay

In 1964, Shapiro measured the decreased velocity of light in a gravitational field, as observed by the speed of a light signal diminishing when passing the gravitational field of the Sun. Shapiro's result is understood by today's physics as a "gravitational time delay" caused by spacetime dilation, which increases the path length. According to bijective research methodology, where every element in the model has the exact correspondent element in physical reality, this interpretation appears not to be exact as Shapiro did not measure spacetime dilation. In SR, the element of "spacetime dilation" has no bijective correspondence in the physical world as it has never been observed in physics that spacetime or space are dilating. According to bijective research methodology, Shapiro's result should be termed the 'gravitational diminishing of light-speed' caused by the diminished energy density of the SQS. In SQS with a given gravitational field, the energy density of the SQS is diminished that causes a minimal diminishing of the speed of light as defined by the permittivity and permeability of the SQS:

$$c = \frac{1}{\sqrt{\mu_0 \varepsilon_0}},$$

where μ_0 is the magnetic permeability and ε_0 is electric permittivity of the SQS where there is no influence of gravity, and the density of the SQS is at its maximum ρ_{Emax}. In the SQS with a gravity field, the energy density of the SQS decreases and causes minimal diminishing of the permittivity and permeability, which in turn result in the minimal diminishing of the speed of light, as presented by Masanori, "it is known that the speed of light depends on the gravitational potential. In the gravitational fields, the speed of light becomes slow, and time dilation occurs. In this discussion, the permittivity and permeability of free space are assumed to depend on gravity and are variable". Minimal variability of the speed of light caused by a gravity field maintains SR because its first postulate is valid only in space where gravity is absent. The electric permittivity in flat space with no gravity is ε_0, and magnetic permeability in flat with no gravity SQS is μ_0. Following Puthoff, on the surface of stellar object, permittivity and permeability are:

$$\varepsilon = K\varepsilon_0,$$

$$\mu = K\mu_0,$$

where the space dielectric constant K on the surface of a stellar object is:

$$K \approx 1 + \frac{2Gm}{rc^2},$$

with G being the gravitational constant, M is the mass, and r the distance from the origin located at the centre of the mass M. Combining equation above with the equation that describes the mass-energy equivalence principle we can write following equation:

$$K \approx 1 + \frac{2G\left(\rho_{Emax} - \rho_{Emin}\right)V}{rc^4}$$

which shows the SQS dielectric constant depends on the variable energy density of SQS. In this sense, a diminished energy density of the SQS on the surface of a given stellar object increases permittivity and permeability of the SQS which, in turn, minimally decreases the velocity of light as:

$$c = \frac{1}{\sqrt{\varepsilon\mu}},$$

where ε is the electric permittivity and μ is its magnetic permeability of the SQS where there is gravitational field. From this, it follows that the Shapiro gravitational time dilation has its origin in the diminished energy density of the

SQS near the stellar objects, which increases the dielectric constant K of the SQS and this minimally decreases the velocity of light. In other words: diminishing SQS energy density $\rightarrow$ increases the dielectric constant $\rightarrow$ increases the electric permittivity of the SQS$\rightarrow$ increases the magnetic permeability of the SQS$\rightarrow$ decreases the velocity of light.

The classic textbook explanation of the Shapiro experiment is that in stronger gravity, time, as the fourth physical dimension of space, dilates causing light to need more time to reach the point B from point A in a spacetime that acts as the fundamental arena of the universe. Bijective research methodology requires an exact explanation where the velocity of light is minimally diminishing in a gravity field due to a diminished energy density of the SQS.

Doppler effect proves the second postulate of SR, which states that "the speed of light c is a constant, independent of the relative motion of the source." The observer exists in SQS, and a photon is the vibration of the same SQS. When the observer moves toward or away from the source of light, they will experience the Doppler effect. With the understanding that the moving observer and the source both exist in the same SQS and that light is the vibration of the SQS, the second SR postulate becomes logical. The observer sees the light with a given frequency coming from the source. When the observer moves away from or closer to the source, the frequency of the light diminishes or increases, respectively.

Gravitational lens

SQS fluctuations bend light, which we refer to as a "gravitational lens", and this bending of light as it passes the Sun is one proof of General Relativity. The SQS fluctuations near the Sun's surface are strongest and push the photons, causing them to bend.

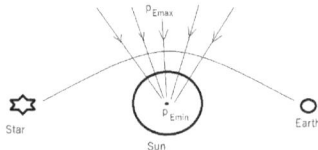

SQS fluctuations bending light around the Sun.

SQS fluctuations bend the photon's trajectory, which we call "gravitational lens". Einstein's formula for the bending of light as it passes the Sun is expressed as:

$$\delta = \frac{4GM_S}{c^2 b},$$

where δ is the angle of deflection, M_S is the mass of the Sun, c is the speed of light, and b is the minimum distance between the trajectory and the centre of the Sun. The mass of the Sun, M_S, can be expressed according to equation of extended mass-energy equivalence principle which we can combine with equation above and we obtain

$$\delta = \frac{4G\,(\rho_{Eman} - \rho_{Emin})V}{c^4 b}.$$

Equation above confirms that SQS fluctuations that carry gravity are directed from SQS where the energy density of the SQS is ρ_{Emax}, towards the energy density of the SQS is ρ_{Emin} in the centre of the Sun. These SQS fluctuations push the photons, causing light deflection. Light passing the Sun is not deflected as a result of the curvature of universal space; measurements by NASA have proven that the universe's space has a Euclidean shape. Light is deflected around gravitational objects, such as the Sun, due to a push from SQS fluctuations.

Gravitational waves are waves of superfluid quantum space

In this book gravitational waves are not represented as "ripples in the fabric of spacetime" because spacetime, as the fundamental arena of the universe, does not exist. Gravitational waves are ripples of the SQS. Gravitational waves change the permittivity and permeability of SQS. As gravitational waves enter the LIGO interferometer, they changed permeability and permittivity of the SQS which minimally changes the speed of light moving in the beams of the interferometer. This minimal change in the speed of light is what is directly measured by LIGO. No direct data exists to confirm that the length of the beams of the interferometer change due to the gravitational waves. How the subtle phenomena of a gravitational wave could shrink or elongate the length of the interferometer beams, which have a solid iron-concrete core, is an unanswered question. The model here presented solves this question through the direct reading of the available data. What is measured by LIGO is the minimal change in light speed due to minimal variations of the permittivity and permeability of the SQS caused by the gravitational wave entering the interferometer.

Our research suggests photons are excitations of superfluid quantum space (also named "superfluid quantum vacuum"). Recent research confirms gravitational waves have a speed close to the speed of the photon since the recent major discovery in physics, the first measurement of gravitational waves, achieved by the LIGO/Virgo collaboration, several events have been registered. In particular, the merging of two neutron stars detected with its electromagnetic counterpart by the FERMI satellite has led to implications of paramount importance. One of them is the speed of gravitational waves now constrained to be extremely close to that of light. In the model presented in this book the photon and gravitational wave are both excitations of the SQS.

Black holes are rejuvenating Systems of the Universe

Active galactic nuclei (AGNs) are throwing in the interstellar space huge jets of energy in the form of elementary particles. The calculation of the energy density of space in the centre of the black hole with the mass of the Sun shows that in the spacetime singularity of such a black hole energy density of space there is so low that atoms become unstable and fall apart in elementary particles. In this sense, AGN is a rejuvenating system of the universe. It transforms its own old matter into fresh energy in the form of jets.

Several pieces of research suggest that superfluid quantum vacuum also named superfluid quantum space (SQS) is the physical origin of the universal space. The idea of spacetime as the fundamental arena of the universe is replaced by the idea that universal space is a type of energy that has superfluid properties. One of these superfluid properties is that every physical object is diminishing the Planck energy density ρ_{EP} of the superfluid quantum space which is the origin of the universal space in its centre exactly for the amount of its mass m and energy E:

$$E = mc^2 = (\rho_{EP} - \rho_{Ec})V,$$

where ρ_{min} is the energy density of the universal space in the centre of the physical object and V is the volume of the object. The "no hair theorem" states that a black hole can be defined by three parameters: mass, electric charge, angular momentum. Considering the variable energy density of universal space, I introduce a new parameter, the "minimal energy density of SQS in the centre of a black hole". By equation above we get:

$$\rho_{Ec} = \rho_{EP} - \frac{mc^2}{V}$$

where ρ_{Ec} is the energy density of SQS in the centre of a black hole, m is the mass of the black hole and V is its volume.

Calculation of the "Schwarzschild energy density"

"Schwarzschild energy density" one can calculate using equation above:

$$\rho_{E.Sch.} = \rho_{EP} - \frac{3m_\odot c^2}{4\pi r_{Sch.}^3},$$

where $m_\odot$ is the mass of the Sun, and its correspondent Schwarzschild radius $r_{Sch.}$ is $3 \cdot 10^3 m$.

$$\rho_{E.Sch.} = 4.64 \cdot 10^{113} Jm^{-3} - 1.58 \cdot 10^{36} Jm^{-3}.$$

When in the centre of the stellar object the value of energy density of SQS ρ_{Ec} is smaller as Schwarzschild energy density, the atoms in the centre become unstable and are falling apart into elementary particles:

$$\rho_{Ec} < \rho_{E.Sch.} \rightarrow atoms\ are\ unstable$$

The Schwarzschild energy density offers a new interpretation of spacetime singularities in the centre of a black hole. Prof Penrose has following view on spacetime singularities: "If, as seems justifiable, actual physical singularities in spacetime are not to be permitted to occur, the conclusion would appear inescapable that inside such a collapsing object at least one of the following holds: (a) Negative local energy occurs. (b) Einstein's equations are violated. (c) The spacetime manifold is incomplete. (d) The concept of spacetime loses its meaning at very high curvature – possible because of quantum phenomena. In fact (a), (b), (c), (d) are somewhat interrelated, the distinction being partly one of attitude of mind". I suggest that spacetime singularity in the centre of black hole indicate that in the centre of a black hole exist critical physical circumstances that we previously defined as "the energy density of SQS e is below the Schwarzschild energy density $\rho_{E.Sch.}$".

According to Newton's Shell theorem in spacetime singularity, gravity force does not tend to the infinite value; it tends to zero. Going inside the black hole at the distance d from the surface towards the centre gravity force on a given object with mass m is diminishing regarding the gravity force on the surface:

$$F_{g1} = \frac{mM_1 G}{r_1} \, ,$$

where m is the mass of a given object, M_1 is the mass of the black hole shell with the radius r_1, and G is gravitational constant. When r_1 is tending to the zero, M_1 is also tending to the zero, and gravity force F_{g1} is also tending to the zero:

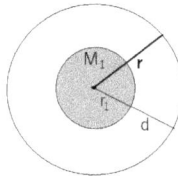

Gravity force inside black hole

In the centre of the black hole, there is no gravity force. The extreme physical circumstance in the centre of the black hole is that the energy density of SQS there is below Schwarzschild energy density. This model is adding to the understanding of the interior of black holes.

Variable energy density of SQS at the distance d from the centre of the stellar objects

The energy density of the universal space ρ_{Ed} at the distance d from the centre of a given stellar object with mass m and radius r is calculated using the equation below:

$$\rho_{Ed} = \rho_{EP} - \frac{3mc^2}{4\pi(r+d)^3}.$$

When d tents towards the infinite, energy density ρ_{Ed} tends towards Planck energy density ρ_{EP}. We will use this formula to calculate the energy density of SQS in the centre of different stellar objects, considering that these objects are non-rotating. In table below there is the comparation of the energy densities of SQS in the centre of the black hole with the mass of the Sun, in the centre of the proton, in the centre of the Moon, Earth, and Sun:

Comparation values of the minimal energy density of space with respect to the centre of indicated objects.

Centre of objects	$\rho_{EP} = 4.64 \cdot 10^{113} Jm^{-3}$
Black hole with mass of the Sun	$\rho_{EP} - 1.58 \cdot 10^{36} Jm^{-3}$
Proton	$\rho_{EP} - 5.45 \cdot 10^{34} Jm^{-3}$
Earth	$\rho_{EP} - 4.97 \cdot 10^{20} Jm^{-3}$
Moon	$\rho_{EP} - 3.01 \cdot 10^{20} Jm^{-3}$
Sun	$\rho_{EP} - 1.27 \cdot 10^{20} Jm^{-3}$

In the centre of a proton, the minimal energy density of SQS is for the order of 10^2 too high for the proton to become a mini black hole as proposed by Hawking. Voyager did not discover mini black holes in interstellar space. The energy density of SQS in the proton centre is lower than in the centre of Sun, Earth and Moon because these stellar objects are made out of atoms where there is a vast empty space between the nucleus and electrons orbits. Proton's mass is very small compared with the mass of the Sun, but it diminishes the energy density of an extremely small area of space compared with that of Sun, that diminishes the energy density of an extremely big area of universal space; that's why the gravity force of the Sun has such a long-range.

We will compare the energy density of SQS of a stationary black hole with the mass of the Sun and energy density of SQS of the Sun at given distances from the centre. Comparation values of the minimal energy density of SQS with respect to the distance by centre of indicated objects.

Sun centre	Black hole centre
$\rho_{EP} - 1.27 \cdot 10^{20} Jm^{-3}$	$\rho_{EP} - 1.58 \cdot 10^{36} Jm^{-3}$
Distance from the centre	Distance from the centre
10^2 km	10^2 km
$\rho_{EP} - 1.27 \cdot 10^{20} Jm^{-3}$	$\rho_{EP} - 3.91 \cdot 10^{31} Jm^{-3}$
10^3 km	10^3 km
$\rho_{EP} - 1.26 \cdot 10^{20} Jm^{-3}$	$\rho_{EP} - 4.24 \cdot 10^{28} Jm^{-3}$
10^4 km	10^4 km
$\rho_{EP} - 1.22 \cdot 10^{20} Jm^{-3}$	$\rho_{EP} - 4.27 \cdot 10^{25} Jm^{-3}$
10^5 km	10^5 km
$\rho_{EP} - 8.48 \cdot 10^{19} Jm^{-3}$	$\rho_{EP} - 4.28 \cdot 10^{22} Jm^{-3}$
10^6 km	10^6 km
$\rho_{EP} - 8.77 \cdot 10^{18} Jm^{-3}$	$\rho_{EP} - 4.28 \cdot 10^{19} Jm^{-3}$
0.1 AU	0.1 AU
$\rho_{EP} - 1.11 \cdot 10^{16} Jm^{-3}$	$\rho_{EP} - 1.28 \cdot 10^{17} Jm^{-3}$
0.5 AU	0.5 AU
$\rho_{EP} - 9.94 \cdot 10^{13} Jm^{-3}$	$\rho_{EP} - 1.02 \cdot 10^{14} Jm^{-3}$
1 AU	1 AU
$\rho_{EP} - 1,26 \cdot 10^{13} Jm^{-3}$	$\rho_{EP} - 1.28 \cdot 10^{13} Jm^{-3}$
10 AU	10 AU
$\rho_{EP} - 1,27 \cdot 10^{10} Jm^{-3}$	$\rho_{EP} - 1.28 \cdot 10^{10} Jm^{-3}$

Going from the centre of the black hole, the energy density of the SQS is increasing at a much higher rate than going away from the centre of the Sun. At the distance of 1 AU from the centre of both stellar objects, the energy density of SQS is at the same rate and is increasing by the same values with the increase of the distance coming closer to the Planck energy density that characteristic for the intergalactic space.

The quantum mechanism of AGNs' jets

In the centre of black holes, atoms are transforming back into elementary particles. This creates enormous pressure and if gravity pressure of the black hole is not big enough, such a black hole explodes in a supernova. When the black hole gravity pressure is strong enough, as it is the case for example with the black hole in the quasar SMSSJ215728.21–360215.1 which has about $(3.4 \pm 0.6) \cdot 10^{10}$ $M_\odot$, the transformation of matter into elementary particles creates the explosion that opens the hole in the direction of the rotational axis.

Cross-section of a black hole in the centre of the quasar SMSSJ215728.21–360215.1

Through this hole, in the direction of rotation, the black hole is throwing a jet of elementary particles into the intergalactic SQS.

Jets of a black hole in the center of a galaxy (with the permission of the European Southern Observatory (ESO)

Centres of AGN's where energy density of SQS is lower than Sch. energy density are mechanisms where the matter falls apart into elementary particles and forms jests. I give in this book a solution to the mystery of jets production following Einstein's idea that matter can be transformed into energy and vice versa. These jets are building material for new stars formation; black holes are then rejuvenating systems of the universe: "old" matter is transformed into "fresh" energy in the form of AGNs jets.

The law of energy conservation requires that AGN's jets must have some physical sources. It is shown in this book that these jets are originated in the process of matter falling apart in the centres of AGNs, where there are according to Penrose model spacetime singularities and energy density of SQS is below the Schwarzschild energy density.

Multiverse in Dynamic Equilibrium

Big Bang cosmology is problematic because of the hypothetical beginning that is not in accord with the conservation of energy. Further, it is based on interpretation of astronomical data that is questionable. CMB is not direct proof of the existence of the recombination period in some remote physical past. Cosmological redshift can be seen as the "tired light effect" proposed by Zwicky. On the basis of direct reading of astronomical data, here we introduce a model of the universe which predicts that in AGNs matter is transforming back into the elementary particles in the form of huge jets that are throwing elementary particles into the intergalactic space and so creating "fresh material" for new stars formation. This process occurring in AGNs, invoked by our model, has no beginning, it is in permanent dynamic equilibrium.

In respect to the results of our research and Rovelli's research on time, we replaced the spacetime model with the superfluid quantum space model. "Superfluid quantum space (SQS) has a general n-dimensional complex structure $\mathbb{C}^n$. Every point of $\mathbb{C}^n$ has complex coordinates:

$$z_i = x_i + i\, y_i \,.$$

(x_i, y_i) $(i = 1, \dots, n)$ is an ordered n-tuple of real numbers $((x_i, y_i) \in \mathbb{R}^n)$; for the purpose of this paper, we consider its subset $\mathbb{C}^4$ where all elementary particles are different structures of $\mathbb{C}^4$ SQS and have four complex dimensions z_i.

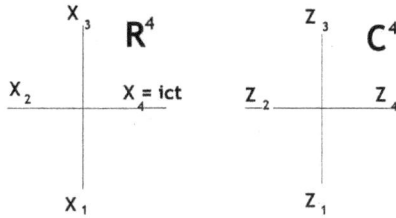

In $\mathbb{C}^4$SQS there is no temporal dimension.

In the cosmology model presented in this book, time does not run independently apart from the change. Time is merely the duration of change. No change in $\mathbb{C}^4$SQS would mean no time. This model is in perfect accord with experimental physics where we measure with clocks the duration of material change that is time. In this sense $\mathbb{C}^4$SQS is timeless, or we say "time-invariant".

Time as duration of changes running in time-invariant SQS also solves the "four-vector" puzzle. In special relativity, the four-vector is introduced in order to unify spacetime coordinates x, y, z, and t into a single entity. The length of this four-vector, called the spacetime interval, is shown to be invariant, which means the same for all observers: $A = (A^0, A^1, A^2, A^3)$, where A^0 as a temporal coordinate is $A^0 = ct$. The so-called "temporal coordinate" is a product of time t as the duration of motion and light speed c. The four-vector can be positive or negative and depends on the direction of motion in future or in past:

$$d\tau = \pm\sqrt{dx^\mu dx_\mu}.$$

where τ is proper time. The idea of motion into past or into future is questionable because it leads to the logical inconsistency where the sum of positive four-vector and negative four-vector is zero:

$$\sqrt{dx^\mu dx_\mu} + \left(-\sqrt{dx^\mu dx_\mu}\right) = 0.$$

This means that the value of the spacetime interval in the Minkowski manifold from A to B and back from B to A is zero which seems wrong. The idea that a given physical object can move in the future or in the past will be re-examined. In experimental physics, we measure with clocks the duration of motion in space. We do not have any experimental evidence that a given physical

44

object is moving in the direction from the past towards the future. In $\mathbb{C}^4$ space there is no past, and there is no future. A given physical object can move only in a $\mathbb{C}^4$ space and not in time that is the duration of motion. The value of the four-vector $A = (A^1, A^2, A^3, A^4)$ in a $\mathbb{C}^4$ space is always positive. There is no negative time $-t$ and the negative four-vector puzzle is solved.

In the 20th century, the idea of moving back in time was widely accepted. Feynman has defined positron as the electron that is moving backward in time. Time was meant to be the physical reality in which elementary particles move; we do not have a single data that would support this idea. With clocks we measure the duration of motion in space, it is time to abandon the idea of time being the 4th dimension of space. Instead, we developed a $\mathbb{C}^4$ space where time is the duration of change.

Gödel development of Einstein field equations of general relativity shows that they lead to the contradiction, namely, one could move back in time and kill his grandfather and so he could not be born. By 1949, Gödel had produced a remarkable proof: "In any universe described by the Theory of Relativity, time cannot exist." He understood that his development of General Relativity proves that time has no physical existence and nobody can travel in time. Still today he is misunderstood by thinking that his work is proving that time travel is possible. Nobody can travel in time because time is not 4th dimension of universal space. The introduction of the $\mathbb{C}^n$SQS as the fundamental arena of the universe where time is the duration of motion resolves the contradiction of "motion in time" and is an important element of cosmology progress.

Cosmological redshift is "tired" light effect

The redshift of the light coming from distant galaxies is today understood as the experimental proof of the universal space expansion. We do not have a theoretical model with mathematical evaluations in scientific literature that exactly predict how the light would behave when moving in the opposite direction of expanding space. This is a serious inconvenience and a puzzle that needs to be solved. The Doppler effect is observed on Earth's surface and Earth is moving around the Sun in the stationary space.

Recent research suggests that the $\mathbb{C}^4$SQS has the value of Planck energy density ρ_{EP}. The gravitational constant G can be expressed with Planck energy density ρ_{EP} and Planck time t_P as:

$$G = \frac{c^2}{\rho_{EP}\, t_P^2}.$$

If the universe would expand, the energy density of the $\mathbb{C}^4$SQS would diminish and consequently the gravitational constant would increase. The gravitational constant was measured first back in 1798 by Henry Cavendish. Since then, the value of gravitational constant is stable, meaning that the density of $\mathbb{C}^4$SQS is also stable. This is suggesting that the universe is not expanding.

Not only the gravitational constant, also the magnetic permeability μ_0 and the electric permittivity ε_0 of the $\mathbb{C}^4$SQS are defined by its energy density. The increase and decrease of the energy density of the $\mathbb{C}^4$SQS would be a cause for the change of magnetic permeability μ_0 and electric permittivity ε_0 and would consequently change the light speed. This last was exactly measured by English astronomer James Bradley back in 1729. The constancy of μ_0, ε_0 and light speed is suggesting that the energy density of the $\mathbb{C}^4$SQS is constant and that universe is not expanding.

Stephen Hawking has predicted that the universe started by the mathematical point. Back in 2014, NASA has measured with the 0,4% of error that the universal space has Euclidean shape by measurement of the sum of angles between three stellar objects and getting 180°: "Thus, the universe was known to be flat to within about 15% accuracy prior to the WMAP results. WMAP has confirmed this result with very high accuracy and precision. We now know that the universe is flat with only a 0.4% margin of error. This suggests that the Universe is infinite in extent; however, since the Universe has a finite age, we can only observe a finite volume of the Universe. All we can truly conclude is that the Universe is much larger than the volume we can directly observe". NASSA report confirms that the universal space can be considered infinite in its volume. On the question how a mathematical point could extend into infinite space of the universe has no answer; we know in mathematics that the mathematical point is dimensionless and cannot be transformed into a given volume.

In FLWR metrics the density parameter Ω ultimately governs whether the curvature is: negative ($\Omega < 1$), positive ($\Omega > 1$), flat ($\Omega = 1$). When density parameter is Ω is 1 in the FLWR metrics universal space has Euclidean shape. In our model, the value 1 of the density parameter Ω is related to the Planck energy density of intergalactic space. In every single point of the universal space, the value of the density parameter Ω is unchanged because in the centre of a given physical object the energy density of superfluid quantum space - $\mathbb{C}^4$SQS is diminishing exactly for the amount of its mass m and energy E accordingly to the equation below:

$$E = mc^2 = (\rho_{EP} - \rho_{Emin})V,$$

where ρ_{Emin} is the energy density of the $\mathbb{C}^4$SQS in the centre of the physical object and V is the volume of the object. This means that the density parameter Ω has the same value in the centre of a black hole and in the intergalactic space.

Considering that density parameter Ω is 1, the only possible future scenario of the universe is Big Rip where all massive objects will have been ripped apart. In Big Bang cosmology we have to invoke "phantom" moments: According to Hawking universe has started from mathematical point and according to Big Rip scenario galaxies will be ripped apart. The cosmology model presented in this book has no such "phantom" moments, it is based only on astronomical data.

We have a plausible explanation of cosmological redshift. When light is coming to us from remote galaxies, it moves against the space fluctuations which are carrying gravity force. $\mathbb{C}^4$SQS fluctuations are flowing from outer interstellar space where $\mathbb{C}^4$SQS has maximum energy density towards lower energy density of $\mathbb{C}^4$SQS in the centre of stellar objects; these $\mathbb{C}^4$SQS fluctuations are carrying gravity force. Light from distant galaxies is moving in the opposite direction of these space fluctuation and is that why losing some of its energy. The result is the cosmological redshift. Swiss astronomer Zwicky has named this effect "tired light effect".

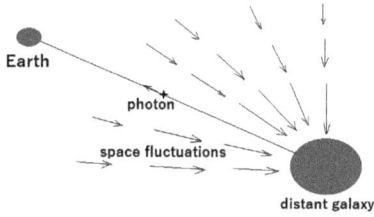

Earth
photon
space fluctuations
distant galaxy

Light is losing some of its energy when moving in the opposite direction of the space fluctuations that carry gravity.

Theory of vector gravity proposed by Anatoly A. Svidzinsky is a model that supports the reinterpretation of gravitational redshift: "Similarly to general relativity, vector gravity postulates that the gravitational field is coupled to matter through a metric tensor f_{ik} which is, however, not an independent variable but rather a functional of the vector gravitational field. In particular, action for a point particle with mass m moving in the gravitational field reads:

$$S_{matter} = -mc \int \sqrt{f_{ik} \, dx^i dx^k}$$

where c is the speed of light. Action has the same form as in general relativity, however, the tensor gravitational field g_{ik} of general relativity is now replaced with the equivalent metric f_{ik} (f_{ik} is a tensor under general coordinate transformations)".

Our model provides the physical origin of vector gravity that is in the $\mathbb{C}^4$SQS quantum fluctuations that are directed from the higher energy density of $\mathbb{C}^4$SQS towards the lover density of $\mathbb{C}^4$SQS. These fluctuations interact with photons to diminish their frequency, which is referred to as 'gravitational redshift.' When light from distant galaxies reaches the Earth, its frequency is lower. On its path to Earth, light loses some of its energy because it is moving against the $\mathbb{C}^4$SQS fluctuations that points toward the direction of galaxies, so that

$$E_{photon.Earth} = E_{photon.galaxy} - \Delta E,$$

48

where $E_{photon.galaxy}$ is the energy of the photon at the galaxy, $E_{photon.Earth}$ is the energy of the arrived photon at the Earth, and ΔE is the loss of energy due to the fluctuations of the $\mathbb{C}^4$SQS,

$$\Delta E = h\Delta\nu,$$

where h is Planck's constant and $\Delta\nu$ is the decrease of the photon frequency due to $\mathbb{C}^4$SQS fluctuations. $\mathbb{C}^4$SQS quantum fluctuation in distant galaxies in the direction from ρ_{Emax} (outer space) towards ρ_{Emin} (centre of galaxy) are the physical origin of so called "tired light" model of astronomer Fritz Zwicky. What we call "cosmological redshift" is "tired light effect".

Recent research of Tiguntsev is confirming that cosmological redshift has its origin in the gravitational field of galaxies from which light is reaching the Earth. When light is moving from the galaxy in the opposite direction of gravity it has a minimal diminishment of velocity. This causes the loss of frequency and consequently the redshift effect.

NASA's discovery means the end of expansion model of the universe and the end of inflation model

NASA has measured back in 2014 that universal space is flat, it has a Euclidean shape. In FLWR metric the density parameter Ω ultimately governs whether the curvature is: negative ($\Omega < 0$), positive ($\Omega > 0$), flat ($\Omega = 0$). When the density parameter is Ω is 1 in the FLWR metric universal space has a Euclidean shape and FLWR metrics predict that such a space can expand. This is against the metrics of Euclidean geometry where the distance between two points is always constant. In an n-dimensional Euclidean space the distance d between point p and point q is calculated accordingly:

$$d\sigma = \left(\sum_{i=1}^{n}(p_i - q_i)^2 \right)^{\frac{1}{2}}.$$

We do not have any possibility in the frame of Euclidean geometry that the distance d would be changed. We cannot expand or shrink Euclidean space. This means that such a space is homogeneous and isotropic. But in spherical geometry or hyperbolic geometry, the situation is different. For example, in a spherical geometry considered as a model for universe, all relative distances increase at a rate proportional to their magnitudes. When using spherical geometry universe is closed. In the hyperbolic case which is an open universe, when the radial coordinate increases away from the origin, the circumferences increase more rapidly with proper radius. One can observe the differences between these three types of spaces from the following:

$$d\sigma^2 = \frac{\sum_{i=1}^{n}(p_i - q_i)^2}{\left(1 + \frac{k}{4}\left(\sum_{i=1}^{n}(p_i - q_i)^2\right)\right)^2}$$

1. Flat space: k=0
2. Spherical space: k > 0
3. Hyperbolic space: k < 0.

Equation above confirms that when $k = 0$, the distance d cannot increase or decrease. The idea that universal space has been inflating immediately after the hypothetical explosion is inspired by the fact that in mathematics, we can increase the radius of the Riemann manifold, and its volume will increase. We have shown in the previous section that we cannot apply Riemann geometry in cosmology because universal space has a Euclidean shape and cannot expand. We do not have a single direct measurement that would prove that universal space is expanding. The idea that universal space could expand has no mathematical basis and has no support in astronomical observations. We have shown in the previous chapter that the Mössbauer effect is a direct proof of the cosmological gravitational redshift.

Back in 2011 Steinhardt published an article in *Scientific American* questioning if inflation is a flawed model: "Is the theory at the heart of modern cosmology deeply flawed?". In his article he did not give final conclusions. He pointed out that the inflation model has some unbridgeable problems that seems are no solvable. Back in 2017 Steinhardt published together with Anna Ijjas and Abraham Loeb another article in *Scientific American* titled "Cosmic Inflation Theory Faces Challenges - The latest astrophysical measurements, combined with theoretical problems, cast doubt on the long-cherished inflationary theory of the early cosmos and suggest we need new ideas". The three authors question the dominant idea of the inflation, the fact that the early cosmos underwent an extremely rapid expansion, suggesting the necessity to consider other scenarios, and in particular the possibility that our universe began with a bounce from a previously contracting cosmos. Their article has opened a feverish debate among world-leading cosmologists. For example, Cornellussen writes in a 2017 *Physics*

Today paper: "The trio's aggressive reappraisal of a scientific consensus inspired an energetic rebuttal, also in *Scientific American*, from 33 prominent physicists, including four Nobel laureates".

We are proposing in our model a new way of solving the problems of the inflation model and also other problems of Big Bang cosmology. We suggest here a cosmological model that will be based on the direct reading of obtained data. Mössbauer effect is directly observed and measured. It confirms that light when moving in the opposite direction of gravity force diminishes its frequency. This is the so-called "gravitational redshift". Cosmological redshift has the same physical origin. When light is pulling out of the strong gravitational fields of distant galaxies their frequency diminishes. This is the manner in which in our model the idea of inflation can be avoided and abandoned. According to our model, universal space is flat, of Euclid nature, and cannot expand; in the light of the Mössbauer effect, when light moves in the opposite direction of the gravitational fields of galaxies their frequencies diminishes.

Alan Guth's view is that universe run in some physical time. With the Big Bang this physical time has entered into existence. How this has happened we do not know yet: "There is much evidence that at earlier times the universe underwent inflation, but the details of how and when inflation happened are still far from certain. There is even more uncertainty about what happened before inflation, and how inflation began. I will describe the possibility of "eternal" inflation, which proposes that our universe evolved from an infinite tree of inflationary spacetime. Most likely, however, inflation can be eternal only into the future, but still must have a beginning. In the same article Guth has continued: "Since inflation is eternal into the future, it is natural to ask if it might also be eternal into the past. The explicit models that have been constructed are eternal only into the future and not into the past, but that does not show whether or not is possible for inflation to be eternal into the past". Guth sees universe running in some physical time that we show is non-existent. His speculations about eternal inflation into future and possible eternal inflation from the past are strictly theoretical and have no experimental evidence.

Guth and the co-authors admitted that inflation model is not self-consistent: "Thus inflationary models require physics other than inflation to describe the past boundary of the inflating region of spacetime". Cosmology model presented in this book is self-consistent.

Guth' way of incorporating gravity in his inflation model is not convincing: "The expansion of the universe may be described by introducing a time-dependent "scale factor," $a(t)$, with the separation between any two objects in the universe being proportional to $a(t)$. Einstein's equations prescribe how this scale factor will evolve over time, t. The rate of acceleration is proportional to the density of mass-energy in the universe, p, plus three times its pressure, p: $\frac{d^2a}{dt^2} = -\frac{4\pi G(\rho+3p)a}{3}$, where G is Newton's gravitational constant (and we use units for which the speed of light $c = 1$). The minus sign is important: ordinary matter under ordinary circumstances has both positive mass-energy density and positive (or zero) pressure, so that $(\rho + 3p) > 0$. In this case, gravity acts as we would expect it to: All of the matter in the universe tends to attract all of the other matter, causing the expansion of the universe as a whole to slow down". By adding the negative mathematical sign in the formula gravity in the universe will not change. In our model gravity cannot be seen as positive or negative in the mathematical sense. Gravity is the result of the diminished energy density of $\mathbb{C}^4$SQS in the centre of a given physical object.

The radius of the mapped universe measured on the basis of astronomic observations is about $4,4 \cdot 10^{26}\ m$. The age of the universe is about $4,35 \cdot 10^{17}s$. According to these data the universe should expand with a velocity of $1,011 \cdot 10^9\ m/s$, that is about 3,3 light speed to reach the mapped size of the universe. The idea that the universe could expand with the average velocity of a 3,3-time of light speed seems unacceptable; we do not have a single theory in physics that would predict such a velocity. The discrepancy between the measured mapped universe and the hypothetical size and expansion of the universe is a big unresolved question of the Big Bang cosmology model.

Hubble law and Doppler effect in an expanding space

Hubble law states that acceleration of the universe increases by the distance:

$$v = H_0 D,$$

where v is the velocity typically expressed in kms^{-1}, H_0 is Hubble constant and D is the distance of the galaxy from the observer measured in megaparsecs (Mpc). One Mpc is $3,261 \cdot 10^6$ light-years. Velocity v of the expansion is defined on the basis of the redshift of a given galaxy. Universal space is expanding and so distances to the galaxies are increasing. The velocity of the galaxies is determined by their redshift that occurs because of Doppler effect. We have shown in section 2 that there is no appropriate mathematical model existing that would describe the Doppler effect in an expanding space. Equation of the Doppler effect is following:

$$f = \left(\frac{c \pm v_r}{c \pm v_s}\right) f_0,$$

f is observed frequency, f_0 is emitted frequency, v_r is the speed of receiver relative to the medium, c is the light speed, and v_s is the speed of the source relative to the medium. Equation above is valid when the medium is at rest. Doppler effect is observed only in the stationary space where electric primitivity ε_0 and magnetic permeability μ_0 of space that define light speed are unchanged. We do not know how the Doppler effect would work in an expanding space where the energy density of the $\mathbb{C}^4$SQS would diminish and electromagnetic properties of space would be changed. Masanori research confirms that gravity influences the electromagnetic properties of space: "It is known that the speed of light depends on the gravitational potential. In the gravitational fields, the speed of light becomes slow, and time dilation occurs. In this discussion, the permittivity and permeability of free space are assumed to depend on gravity and are variable". Applying the Doppler effect in Hubble law without knowing how the expansion of the universe changes electromagnetic properties of expanding space seems unacceptable.

Back in 2019, NASA has reported on universe expansion: "The new estimate of the Hubble constant is 74 kilometers (46 miles) per second per megaparsec. This means that for every 3.3 million light-years farther away a galaxy is from us, it appears to be moving 74 kilometers (46 miles) per second faster, because of the expansion of the universe. The number indicates that the universe is expanding at a 9% faster rate than the prediction of 67 kilometers (41.6 miles) per second per megaparsec, which comes from Planck's observations of the early universe, coupled with our present understanding of the universe".

The average velocity of the universe expansion that is according to the size of the mapped universe and age of the universe 3.3-time of light speed which yield $9.893 \cdot 10^8 ms^{-1}$. Hubble constant is measured to be 74 kilometers per second which yield $7.4 \cdot 10^4 ms^{-1}$. According to the value of the Hubble constant universe should be much smaller. This is the second weak point of Hubble law.

Hubble law predicts the existence of the Hubble sphere, a spherical region of the observable universe beyond which objects recede at a rate greater than the speed of light due to the expansion of the universe. How galaxies could have velocity higher than light speed is also an unanswered question of Hubble law. Research published in 2013 has confirmed that photons form matter. This means that every physical object accelerated to the light speed would turn into light. No physical object can move with light speed. Only photons can move with light speed. The Hubble sphere model is suggesting that beyond the Hubble sphere there are only photons in the universe and that they move faster than light speed. This seems unacceptable.

Wendy Freedman has discovered that measurements of the Hubble constant based on the astrophysics of stars and CMB have a 10% of the discrepancy: "It is certainly worth noting that the local measurement of H_0 is based on the astrophysics of stars, and the CMB results are based on the physics of the early universe: the results are entirely independent of each other. 13.8 billion years of evolution of the universe has occurred since the surface of last scattering of the CMB and the present day, and yet the two measures agree to within 10%. Viewed from a historical perspective, the agreement is actually rather remarkable". Freedman is pointing out that this discrepancy is signalling the cosmology beyond the standard model: "Over the past 15 years, measurements of the fluctuations in the temperature of the remnant radiation from the Big Bang have provided a relatively new means of estimating the value of the Hubble constant. This very different approach has led us to an interesting crossroads, yielding a lower derived value of H_0. If this discrepancy persists in the face of newer and higher precision and accuracy data, it may be signaling that there is new physics to be discovered beyond the current standard model of cosmology".

Lucas Lombriser has tried to solve this discrepancy with the proposal of a higher local density of matter: "A significant tension has become manifest between the current expansion rate of our Universe measured from the cosmic microwave background by the *Planck* satellite and from local distance probes, which has prompted for interpretations of that as evidence of new physics. Within conventional cosmology a likely source of this discrepancy is identified here as a matter density fluctuation around the cosmic average of the 40 Mpc environment in which the calibration of Supernovae Type Ia separations with Cepheids and nearby absolute distance anchors is performed". Lucas Lombriser is applying in his calculations FLWR metrics as that the cosmic bubble of the 40 Mpc environment would be a universe apart. This seems unacceptable, you cannot take a part of the universe out of the context and calculate the local expansion rate. It makes no sense; if we imagine Big Bang as an initial explosion there is no way according to the known physics that some parts of the explosion would have a different rate of expansion. L. Freedman's proposal of searching beyond Big Bang cosmology deserves serious consideration.

Super-fluid quantum space, dark energy and dark matter

Super-fluid four-dimensional complex quantum space $\mathbb{C}^4$SQS is the primordial energy of the universe. According to the law of energy conservation, this energy cannot be created and cannot be destroyed. Every physical object with the mass m is diminishing the energy density of $\mathbb{C}^4$SQS in its centre exactly for the amount of its energy E. Variable energy density of $\mathbb{C}^4$SQS is generating inertial mass m_i and gravitational mass m_g of a given physical object: "From the macro to the microscale, it holds that a given physical object is interacting with the $\mathbb{C}^4$SQS in which is existing; the result of this interaction are the inertial mass m_i and the gravitational mass m_g:

$$m_i = m_g = \frac{(\rho_{EP} - \rho_{Emin})V}{c^2}.$$

The inertial mass of a given physical object is not its rest mass, it is the result of the interaction of rest mass with the $\mathbb{C}^4$SQS. Gravity force between two physical objects is as follows:

$$F_g = \frac{m1_g m2_g G}{r^2},$$

where $m1_g$ is the inertial mass of the first object and $m2_g$ is the inertial mass of the second object:

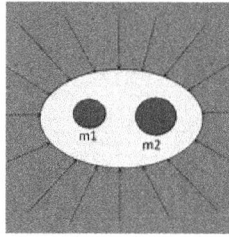

Gravity for acts from outer space towards the centre of physical objects

Higher pressure of outer $\mathbb{C}^4$SQS is pushing together physical objects. $\mathbb{C}^4$SQS is the "unknown fluid" of the universe. It cannot have negative pressure as suggested in by Rajendra Prasad and his research fellows: "This acceleration in the universe may driven by an exotic type of unknown fluid that have positive energy density and huge negative pressure. This fluid is usually known as Dark Enegry (DE) but its nature is still unknown. The most suitable candidate of this DE is the Λ. However, there is a huge dissimilarity in the value of Λ predicted by observations and particle physics ground that leads tuning problem". Negative pressure of the "unknown fluid" is a theoretical proposal that was never observed in experimental physics and cannot be taken as a stable ground to build cosmology.

Dark energy, also named "unknown fluid is 68% of the energy in the universe. About 5% of the energy in the universe is in the form of visible matter, while about 27% is in the form of dark matter and about 68% of the energy of the universe is in the form of dark energy. Since the 1980s, the dominant paradigm for the nature of dark matter has been that of the weakly interacting massive particle (WIMP).

In our model, the energy of the $\mathbb{C}^4$SQS is the dark energy; the idea that universal space is empty and dark energy is hidden somewhere in the space seems wrong. $\mathbb{C}^4$SQS model offers the solution for the discrepancy between measured and theoretically valued cosmological constant. The measured value of cosmological constant $\Lambda = 5.96 \cdot 10^{-27}$ kg/m^3 is different from its calculated value following the Planck metrics for the magnitude of 10^{123}. This discrepancy is an unsolved subject of physics for decades. Regarding the suggested energy density of space proposed in this article, we are defending our proposal by the fact that the gravitational constant G is obtained by measurement and is expressed by the Planck energy density ρ_{EP} and the Planck time t_P as:

$$G = \frac{c^2}{\rho_{EP}\, t_P^2}.$$

This means that the Planck energy density ρ_{EP} reflects the real energy density of a 4-D universal space. In the absence of stellar objects, the energy density of the universal space has a value of Planck energy density which is $\rho_{EP} = 4.64 \cdot 10^{113} Jm^{-3}$. Einstein had proposed that universal space is four-dimensional. In his vision time is the 4th dimension of space. In our model also 4th dimension is spatial, time is the duration of the change in time-invariant space.

CMB is the radiation of the existent universal space

We propose also that CMB should not be interpreted as a proof of recombination period. Astronomical observations confirm that the universal space is radiating uniform CMB radiation. The Big Bang model suggests that the CMB radiation is the relic radiation from some remote physical past. The universal space is timeless; no signal can move through some hypothetical physical time; all signals move in the timeless space. The idea that CMB is radiation from some remote physical time is not falsifiable and should be abandoned in the name of cosmology progress. CMB has its source in present time-invariant universal space. Experimental physics is confirming a given signal we can only reach from the existent physical source, and the remote physical past is physically non-existent. Any kind of radiation must have a physical source; the remote past event cannot be this physical source. The proposal that the CMB signal is relic radiation that was created in some remote physical past and is still present is an ad-hoc proposal that was never confirmed by an experiment. The discovery of CMB absolutely does not prove the existence of a recombination period that should be existing around 380000 years after some hypothetical Big Bang. The scientific fact on the basis of observations is that CMB is the radiation of existent universal space that has its physical origin in $\mathbb{C}^4$SQS that is timeless. The CMB has a thermal black body spectrum at a temperature of $2.72548 \pm 0,00057K$.

Big Bang cosmology and Einstein's steady-state cosmology have no answer about matter creation

Alan Gut hypothesis is that the energy of gravity and that of matter have been multiplying in inflation period. The energy of gravity E_g is negative, the energy of matter E_m is positive, their sum is zero and in inflation on the contrary they multiply. We can describe his idea mathematically as follows:

$$nE_m+(-nE_g)=0.$$

Firstly, we never observed negative gravitational energy. Secondly, we never observed that energies are multiplying out of nothing. Gut's idea is against the first law of thermodynamics and is not bijective. There is no logical answer also about where both energies came into existence in the hypothetical inflation. Equation above is mathematically right, but it does not fulfil the test of bijectivity, meaning that it does not correspond to some real process in physical world. The Big Bang model is not falsifiable.

The model of the universe presented in this book is based only on the obtained experimental data, is falsifiable. There are no theoretical speculations as in the case of the Big Bang model. The cosmology model our research group has developed is based only on direct reading of experimental data. Thinking that the gravitational energy could be negative is logically inconsistent, because we never observed to date positive or negative energy in the universe. We know that there are precise conventions on the sign of energy, conventions adopted in all areas of physics, such as thermodynamics (absorbed energy = positive; energy released energy = negative). But these are adopted conventions, no one has ever measured that energy has an associated mathematical sign. This is also in line with the principle of bijectivity introduced in this book. Also, the idea that the energy of the universe is multiplying in the hypothetical inflation is logically inconsistent, because we have no experimental evidence that energy can get multiplied. The inflation is against the first law of thermodynamics.

In the past century, gravity was understood as the force produced directly by the matter, the idea was that universe must be finite. We can read in the article of Sir James Jeans in Nature back in 1943: "If, however, the distribution is uniform throughout the whole of space, then space must be finite; otherwise, it would contain an infinite amount of matter, and the gravitational force from this would be infinite, which is contrary to the fact".

NASA has measured that the universe has Euclidean shape and is infinite. The idea of $\mathbb{C}^4$SQS being infinite does not mean that gravity should be infinite, as suggested by Sir James Jeans. Considering universal space is infinite there is no gravity force between the stellar objects that are on the infinite distance.

The energy of the infinite universe in the form of matter E_m and in the form of superfluid quantum space energy E_{SQS} is infinite:

$$E_m + E_{SQS} = \infty.$$

The human mind can only imagine a finite amount of matter and a finite amount of energy and finite space which is not the case with the universe. The universe is infinite by means of matter, energy, and volume. That's why is opportune we study the universe that is at a finite distance and we predict that the rest of the unobservable universe on the infinite distance is behaving in the same way as our observable universe.

Mass of every physical object in the universe diminishes the energy density of space, the variable energy density of space is carrying gravity that is the fundamental force of the universal dynamics. Defining gravitational energy negative, as done by Hawking and Guth, is questionable; energy is not positive, it is not negative, energy simply is, it cannot be created and it cannot be destroyed, it transforms continuously.

Einstein has proposed on his steady—state theory of the universe that matter is continuously created out of the universal space: "In the final part of the manuscript, Einstein proposes a physical mechanism to allow the density of matter remain constant in a universe of expanding radius - namely, the continuous formation of matter from empty space: "If one considers a physically bounded volume, particles of matter will be continually leaving it. For the density to remain constant, new particles of matter must be continually formed within that volume from space". How the matter is formed out of space Einstein did not explain. Both, Hawking's and Einstein's solution for how matter appears in the universe are pure theoretical speculations. In our model appearance of matter in the universe is not questionable. In AGNs' matter is constantly disintegrating in elementary particles that are fresh energy for matter formation.

Multiverse is in permanent dynamic equilibrium

$\mathbb{C}^n$SQS is multidimensional. All elementary particles are different structures of a $\mathbb{C}^4$SQS. Physical objects are made out of atoms that are three-dimensional. Different layers of $\mathbb{C}^n$SQS are coexisting, they are interwoven. In our view of the multiverse theory, we do not have some parallel universes that are coexisting in some unexplainable way. The universe we perceive and observe is a multiverse. We can only perceive and measure the 3D and the 4D realms of the multiverse. Higher dimensions are not reachable with apparatuses but this does not mean that they are non-existent. Recent research of Hameeda and co-authors is confirming that the idea of the multiverse or "multiple universes" is present in the human culture for ages: "Widely propounded in cosmology, physics, astronomy and hypothesized in philosophical and religious literature, the concept of multiple universes under the names of multiverse, parallel universes, quantum universes or interpenetrating dimensions has been under the debate among the prominent physicists since middle ages". 5D and higher dimensionalities of $\mathbb{C}^n$SQS represent the mathematical model that can describe "hidden variables" of Einstein and "implicate order" of David Bohm.

In our cosmology model the energy density of $\mathbb{C}^4$SQS in interstellar space has a value of Planck energy density $\rho_{EP} = 4.64 \cdot 10^{113} Jm^{-3}$. Every stellar object is diminishing energy density of the 4th dimension of $\mathbb{C}^4$SQS in its centre exactly for the amount of its mass m and energy. Let's see the values of $\mathbb{C}^4$SQS energy density in the centre of some stellar objects on the table below:

Comparation values of the energy density of space with respect to the centre of indicated objects.

Centre of objects	$\rho_{EP} = 4.64 \cdot 10^{113} Jm^{-3}$
Black hole with mass of the Sun	$\rho_{EP} - 1.58 \cdot 10^{36} Jm^{-3}$
Earth	$\rho_{EP} - 4.94 \cdot 10^{20} Jm^{-3}$
Moon	$\rho_{EP} - 3.00 \cdot 10^{20} Jm^{-3}$
Sun	$\rho_{EP} - 1.26 \cdot 10^{20} Jm^{-3}$

In the centre of a black hole with the mass of the Sun and corresponded Schwarzschild radius $r_{Sch} = 3 \cdot 10^3 m$, the minimal energy density of $\mathbb{C}^4$SQS is for the order of 10^{16} lower than in the centre of the Sun. Because of this special

60

physical circumstance atoms become unstable. In the huge black holes in the centre of AGNs matter is falling apart into elementary particles that form jets. Black holes in the centre of galaxies are throwing these jets into intergalactic space. These jets are fresh energy for new stars formation; black holes are rejuvenating systems of the universe. AGNs in the centres of galaxies are keeping entropy of the universe constant: "old" matter is transformed into "fresh" energy in the form of elementary particles .

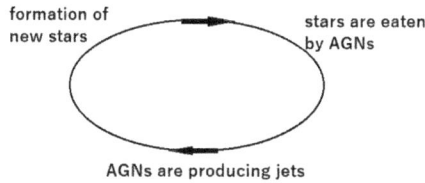

Energy circulation in the universe is permanent.

This process did not start and will never end, it is in permanent dynamic equilibrium. There was no creation of the energy of the universe and there will be no destruction of the energy. An increase of matter entropy in the universe is only a partial process that does not influence the total entropy of the universe that is constant. In AGN-s the universe is rejuvenating itself.

Big Bang cosmology model timeline seems wrong

There is a strong astronomical evidence that the star HD 140283 has an age of 14.27 billion years that is a new difficulty for Big Bang model according to which the age of the universe is calculated by about 13.7 billion years. This astronomical observation is another puzzle Big Bang cosmology cannot solve.

The next problem of the existing big bang timeline is the formation of the galaxies in the early universe, the so-called "early galaxies problem". Several galaxies with such a high redshift are discovered that they should be formed earlier as the big bang model is predicting. Steinhardt and co-authors reported in their article: "We have shown that recent observations of high-redshift galaxies are inconsistent with current theoretical models of galactic assembly. As a general principle, when theory and observation disagree, it is historically best to believe the observational result. However, in this case the observations also rely on untested theoretical assumptions about stellar evolution. Thus, something is wrong, but what?". We suggest that the theoretical assumption of the universe starting with some hypothetical big bang seems wrong.

The next problem of the existing big bang timeline is the discovery of a giant arch behind galaxy cluster IDCS J1426.5+3508 that should accordingly to the big bang cosmology should not exists. Gonzalez and co-authors reported in their article: "Very simply, the arc we have discovered behind IDCS J1426.5+3508 is not predicted to exis. In our cosmological model universe has no "timeline". All stellar objects and formations that we observe do not pose any problem.

Comparing with the big bang cosmology our cosmological model is incorporating the existence of methuselah star HD 140283, the existence of giant arch behind galaxy cluster IDCS J1426.5+3508 and is solving the "early galaxy problem". We developed a cosmology model without the beginning of the universe, the problem of creation is solved. Penrose and Gurzadyan's "Conformal cyclic cosmology" (CCC) model also suggest that the universe is non-created, eternal, and in the permanent cyclic transformation. CCC cosmology is accepting the inflation period that the "CPT – Symmetric universe" model is denying. CPT model predicts that before the big explosion there was an anti-universe in some negative time. We categorically exclude that universe could exist in some negative time or could exist in some positive time. CCC cosmology model and CPT – Symmetric universe model weak points are that both models predict some events in the past that were never observed directly, their existence is questionable. Multiverse in dynamic equilibrium (MDE) is advanced in the sense it is based only on astronomical observations; it has no theoretical speculations about some past events in some remote physical past. MDE model is based on the astronomical observations of the existing observable universe. In MDE model atoms are 3D structures composed out of elementary particles that are different 4D structures of $\mathbb{C}^4$SQS. 5% of the energy in the universe is in the form of matter that is 3D and 95% is in the form of 4D and higher dimensional layers of $\mathbb{C}^4$SQS; dark energy and represent about 68% of the energy of the universe, the weakly interacting massive particle (WIMP) that represent around 27% of the energy of the universe. In an MDE model, the proportion (5% - 27% - 68%) between ordinary matter, dark matter, and dark energy that is the $\mathbb{C}^4 SQS$ energy is more or less constant. A multiverse is a dynamic system in permanent equilibrium. The transformation of mater into elementary particles in the centre of AGNs is permanent; the multiverse is continuously recreating itself. Multiverse is non-created and eternal.

Bio-cosmology - Multiverse, Life and Consciousness

The evolution of life on the planet Earth is happening primarily in the universe and secondary on the Earth. We will examine in this book evolution of life as the cosmic phenomena. In our model multidimensional time-invariant superfluid quantum space that is the fundamental arena of the universe and represents about 95% of the energy in the universe has stable entropy, it is syntropic. The increase of entropy happens only by about 5% of the energy in the universe that is in the form of matter. The evolution of life in our model is a process of matter organization into living systems that tends to develop towards the constant entropy of the time-invariant multidimensional quantum space. This process runs in the entire universe. The development of life into intelligent organisms is the universal process running throughout the entire universe.

The result of several pieces of research is that the superfluid quantum vacuum also named superfluid quantum space (SQS) is the physical origin of the universal space, the fundamental arena of the universe. Elementary particles proton, electron, and photon are 4-dimensional structures of the $\mathbb{C}^4$SQS and have according to the existing quantum theory almost infinite lifetime. Sbitnev proposal is that elementary particles are different vortex structures of superfluid quantum space. As $\mathbb{C}^4$SQS has stable entropy, proton, electron and photon have stable entropy.

The evolution of life in the universe is an intrinsic tendency of 3D matter to develop into systems (living organisms) that tend to develop towards the constant entropy of $\mathbb{C}^n$SQS.

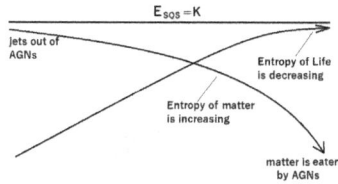

Life is developing towards the constant entropy of $\mathbb{C}^n SQS$.

The entropy of matter in the universe is continuously increasing. The entropy of life is continuously decreasing. The entropy of $S_{\mathbb{C}^n SQS}$ is constant:

$$S_{\mathbb{C}^n SQS} = K.$$

Quantum mechanics of life negentropy

Fritz Popp's and Cohen's research has shown that a living organism has a coherent electromagnetic field that plays an essential role in the organism's function. Electromagnetic fields are carried by the complex four-dimensional superfluid quantum space $\mathbb{C}^4 SQS$. A living organism has an atomic and molecular layer entropic layer and electromagnetic layer that is negentropic. The negentropic layer is the software and the entropic layer is the hardware. Our proposal is that life is an "orchestra" of the higher dimensional layers of $\mathbb{C}^n SQS$. Consciousness in our model is the energy E_c of the photon of the n-dimensional layer of $\mathbb{C}^n SQS$; its frequency tends to the infinite:

$$E_c = v_{\to \infty} h ,$$

where v is photon frequency and h is a Planck constant. Consciousness is governing life via lover dimensional SQS by the pilot photons. Biophotons are studied in detail by Popp and Cohen. It is experimentally proved that photons in integrated photonic devices have a spin. Left spin, we can take like 1, and right spin we can take like 0. When a biophoton is passing the microtubule, it passes the information via its spin. $\mathbb{C}^4 SQS$ photons have 4 bites of the information. They are getting information from higher dimensional SQS photons and are passing it to the microtubule [9]. The research group from China has proved that human high intelligence is involved in the spectral redshift of biophotons activities in the brain.

In our model higher dimensional layers of $\mathbb{C}^n$SQS is the information basis for the development of life. The equation for the increase of information in higher dimensional layers of SQS is following:

$$C_k(n) = \frac{n!}{(r!(n-r)!)} \quad (4)$$

where n is the number of SQS dimensionality, and $r = 3$ because microtubules are 3-dimensional. A four-dimensional biophoton carries 4 bits of information: $[X_1, X_2, X_3]$, $[X_2, X_3, X_4]$, $[X_1, X_2, X_4]$, $[X_1, X_3, X_4]$ and transfers it to the 3D microtubules.

Information density in higher dimensions of $\mathbb{C}^n$SQS.

$\mathbb{C}^4$SQS	4 bit
$\mathbb{C}^5$SQS	10 bit
$\mathbb{C}^6$SQS	20 bit
$\mathbb{C}^7$SQS	35 bit
$\mathbb{C}^8$SQS	56 bit
$\mathbb{C}^9$SQS	84 bit
$\mathbb{C}^{10}$SQS	120 bit
$\mathbb{C}^{100}$SQS	161700 bit
$\mathbb{C}^n$SQS	∞ bit

In n-dimensional SQS the amount of information is infinite. Seems, life and the entire universe are functioning via binary logic and binary transfer of information. That's why we managed the immense development of computers; we discovered the mechanisms of information storage and transfer that are universal. The numbers sequence 4,10,20,35,56,84,120…….is a tetrahedral sequence of numbers, also called triangular pyramidal numbers. It is interesting that several molecules have a tetrahedral structure. Tegmark is proposing that the entire universe is a mathematical structur. Comparing Tegmark's proposal our model is moderate and proposes that the entire three-dimensional universe is built accordingly to the mathematical structures that have their information basis in the higher-dimensional layers of SQS.

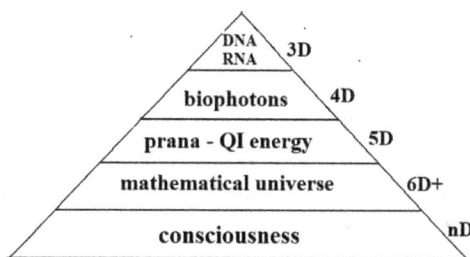

Life-information system in $\mathbb{C}^n$SQS which exists in the entire universe.

Biophotons in a $\mathbb{C}^5$SQS are the physical basis of prana, biphotons in higher dimensional layers are the physical origin of the mathematical universe and finally, biophotons in $\mathbb{C}^n$SQS are the origin of consciousness.

One bit of information in life-information system is a "complex bit", because biophotons are excitations of the complex $\mathbb{C}^n$SQS. One bit of information in artificial intelligence is carried by the electrical current, because in computers one bit means that electrical current has moved in one or other direction. That's why artificial intelligence and alive intelligence will never be compatible. Computers will never have real human emotions and will never be conscious. They will never develop higher cognitive functions that are characteristic for the higher dimensional layers of the human $\mathbb{C}^n$SQS.

Materials and methods

Research done between 1987-90 has confirmed that the presence of the higher dimensional layers of $\mathbb{C}^n$SQS in a living organism minimally increases its weight. Preliminary experiments has been carried out at the Biotechnical faculty, Ljubljana, Slovenia in June 1987. Measurements have been performed on a Mettler Zurich M5 scale. Six test-tubes were filled with three milliliters of a water solution made out of meat and sugar. Four test-tubes were used and a fungus was put into two of the test-tubes. All of test tubes were welded airtight. The weight difference between test-tubes was measured for ten days. After three days of growth, the weight of test-tubes with the fungus increased (on average) 34 micrograms and in last seven days remains unchanged. The experiment was carried out in sterile circumstances and has confirmed that when organic mass turns into an alive mass its weight increases accordingly to:

66

$$Fg_{living.organism} = Fg_{organic.matter} + Fg_{life}$$

The experiment was then carried out in the opposite way. Two test-tubes were filled with 5 grams of Californian worms with distilled water. All of the test-tubes were then welded airtight. The weight difference between test-tubes was measured for 5 hours. At the end of the first hour there was no appreciable difference but at the end of the second and third hour there mass was decreased of 4.5 micrograms on average. This mass then remained stable for the next 2 hours most likely due to there no longer being any living organisms. Experiment was repeated 5 times. The weight loss can be expressed by following equation:

$$Fg_{dead.organism} = Fg_{living.organism} - Fg_{life}$$

Experimental and control test tube back in 1988

These experiments were repeated from August to September of 1988 at the Faculty for Natural Science and Technology, Ljubljana. Two Mettler Zurich scales, type H20T were used in the measurements. A test-tube was filled with 70 grams of live Californian worms and a small testtube was filled with 0.25 ml of 36% water solution of formaldehyde. The control test tube is containing 70 ml of distilled water with a small test tube of formaldehyde inside. Both the test tubes were welded, wiped clean with 70% ethanol, and put into the weighing chamber of the balance. Approximately, one hour was allowed for acclimatization. Later both test-tubes were measured three times at intervals of five minutes. Then the test tubes were turned upside down to spill the solution of formaldehyde and again they were measured seven times at intervals of fifteen minutes. The weight of the test-tube with the worms was found to have increased in the first 3 minutes after the poisoning on average for an average weight of 60 micrograms and it then went down. Fifteen minutes after poisoning, the weight diminished on average by 93.6 micrograms.

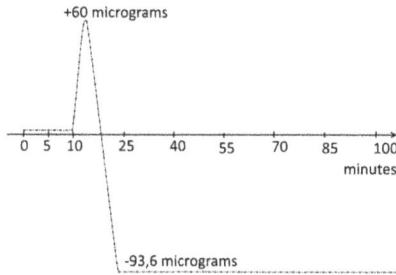

Weight diminishing at the time of worm's death

This last experiment was repeated twelve times. The standard deviation goes to 16 micrograms. The pressure in both test tubes was one atmosphere for the entire duration of the experiment as well as the temperature remaining unchanged. Neither the pressure nor the temperature could have been the cause for the change in the weight. Experiments are preliminary and need to be repeated at least in two different laboratories. We do not encourage researchers to use higher developed animals in this experiment.

In 1997, one of the authors (A. Sorli) published the results of the experiments in the 'Newsletter' nr. 18-19 of Monterey Institute for Study of Alternative Healing Arts, California. On March 3rd 1998, Dr. Shiuji Inomata from Japan informed the editor (S. Savva) that Dr. Kaoru Kavada got similar results using rats as the experimental organism, again in a closed system.

Back in 2019, the experiment with 5 grams of worms was repeated on the high accuracy balance Mettler-Toledo AX107H Comparator. The difference between the two test tubes with the worms and the two control test tubes with destinated water was measured simultaneously. The same results were obtained.

Two experimental and two control test tubes on Mettler-Toledo AX107H Comparator

Today, the interpretation of these experiments is that the mass m of living worms and the mass m of dead worms are the same because in both masses we have the same atoms. Only their molecular composition after poisoning with formaldehyde is different. A living organism has more energy than the same dead organism. Its energy is following:

$$E_{life} = mc^2 + E_{\mathbb{C}^n SQS},$$

where m is the mass of the organism and $E_{\mathbb{C}^n SQS}$ are higher-dimensional energies of $\mathbb{C}^n SQS$ that are present in the living organism. The presence of higher-dimensional energies of $\mathbb{C}^n$SQS in living organism minimally increases its weight.

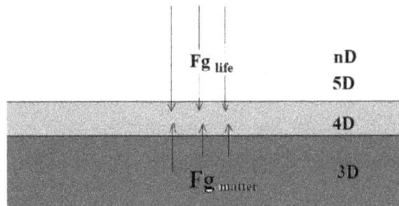

Gravity of matter (Fg_{matter}) and gravity of life (Fg_{life})

The weight of living organisms has two components. The weight of the atoms in the living organism (Fg_{matter}) and the weight of the higher dimensional energies of $\mathbb{C}^4$SQS that are present in the living organism (Fg_{life}).

Discussion on obtained results

The weight difference by humans was first measured by American medical doctor Duncan MacDougall back in 1901. He measured that the weight of the human body after death diminished by about 21.3 grams. Duncan has predicted that the "soul" is leaving the body and so weight is diminishing. In this article, we explained this weight difference in terms of higher dimensional energy layers of superfluid quantum space which presence in the living organism causes the minimal increase of weight.

Several pieces of research are reporting the near-death experience and out-of-the-body experience of dying people. The model of higher-dimensional layers of superfluid quantum space is the theoretical frame that explains these phenomena.

Our model also supports the vibrational theory of DNA which is suggesting that DNS has a kind of "electromagnetic informational duplicate". We suggest that physical DNA is 3D and has its information duplicate in $\mathbb{C}^4 SQS$.

The model of n-dimensional superfluid quantum space also represents the theoretical basis for "DNA phantom effect". DNA has its "electromagnetic information duplicate". When we place the DNA in the experimental tube used in "DNA phantom effect" experiment this duplicate will transfer the information on the $\mathbb{C}^4 SQS$ level. The information will remain there also when DNA is removed.

Herbal drugs cure also on higher dimensional lawyers of $\mathbb{C}^n SQS$ and synthetic drugs are curing only on the 3D molecular level. Herbal drugs have high informational compatibility with the living organism. Synthetic drugs have low informational compatibility that is restricted only on the 3D molecular level. This is one of the causes of the side effects of synthetic drugs that natural drugs do not have has reported Karimi and the co-authors: "About 8% of hospital admissions in the United States of America are due to adverse or side effects of synthetic drugs. Approximately 100,000 people each year die due to these toxicities. It means that the killed people in the U.S. by pharmaceutical drugs are at least three times more than the killed by drunken drivers. Each year also thousands of people die from supposedly "safe" over-the-counter drugs. Deaths or hospitalizations due to herbs are so rare that they are hard to find. Even, the National Poison

Control Centres of the United States does not have a category in their database for side or adverse reactions to herbs". The side effects of mRNA therapy are from the perspective of "informational compatibility" is unknown. How much side effects will cause the synthetic mRNA is un unexplored subject. Shibo Jiang is warning the scientific community: Don't rush to deploy COVID-19 vaccines and drugs without sufficient safety guarantees". Our model of $\mathbb{C}^n SQS$ is extending the understanding of the side effects of synthetical drugs.

Slawinski has measured bio-photon radiation at the time of death of the organism and increases from 10 to 100 times. This confirms that at the time of death the four-dimensional layer of $\mathbb{C}^n SQS$ that represents the coherent electromagnetic field discover by Popp and Cohen is falling apart and this causes increased bio-photons radiation.

Anaesthesia is temporarily breaking the bond between the 3D molecular level of the living organism and its higher dimensional layers and so also with consciousness. That's why the person that is under anaesthesia becomes unconscious. The experiments done on fruit flies *Drosophila confirms that* anaesthesia changes the spin of the electrons in the cells. Turin and co-authors reported in their article:"We propose that anesthetics perturb electron currents in cells and describe electronic structure calculations on anesthetic–protein interactions that are consistent with this mechanism and account for hitherto unexplained features of general anesthetic pharmacology". We suggest that the change of the electron currents perturbation causes that the information line between the 3D part of the organism and the higher dimensional part of the organism (psyche) is temporarily broken.

In our cosmology model, gravity force is the result of the diminished energy density of $\mathbb{C}^4 SQS$ because of the presence of the physical object. The presence of higher dimensional layers of $\mathbb{C}^n SQS$ in living organisms diminishes the energy density of $\mathbb{C}^4 SQS$ and so gravity force is minimally increased. We could say that so-called "subtle energies" as "Prana" or "'QI" energy and consciousness have some minimal weight. Mechanistic science is strictly denying the existence of a reality that reaches beyond electromagnetism. We think this approach will not give us any progress. As Nicola Tesla said: "The day science begins to study non-physical phenomena, it will make more progress in one decade than in all the previous centuries of its existence".

Evolution of life, order, disorder and randomness

We take a "fair coin" and we throw it. We have a 50% possibility to get the "upper side" and a 50% possibility to get a "downside" of the coin. We take a "fair dice" with six numbers. We throw it and we have a 16.66 % possibility to get number six.

We take two fair dices, we place them on the plate so that they both have number six on the upper surface and we throw them. We use the equation (4) to get the number of possibilities. Number n is 12 because we have 6+6 surfaces, and number k is 2. Throwing two dices we can get 66 different combinations. This means that the possibility that both dices will have at next throw number six on the upper surface is 1.56%. Now we take 10 dices and we place them so that all have the number six on the upper surface. Number n is 60, and number k is 10. At the next throw, we have 75394027566 different possibilities. Possibility that all dices will have at next throw number six on upper surface is $1.326 \cdot 10^{-9}$ %. At 100 dices number n is 600, and number k is 100. At the next throw, we have $1.111 \cdot 10^{116}$ different possibilities. Possibility that all dices will have at next throw number six on upper surface is $9 \cdot 10^{-115}$ %.

Random hitting of dices increases the disorder of the system. A living organism's order is extremely bigger than the order of the system of 100 dices. Life is regarding the geological environment extremely high organized system. Longo and Montévil have proposed that randomness increases order in biological evolution. The calculations above confirm that the idea that randomness is the cause of biological evolution seems unacceptable.

Penrose and Hameroff have proposed consciousness as the core of life evolution. They have created orchestrated objective reduction theory (Orch OR), which sees life and consciousness as phenomena that are deeply related to the structures of the universe: "The DP (Diósi–Penrose) form of OR is related to the fundamentals of quantum mechanics and spacetime geometry, so Orch OR suggests that there is a connection between the brain's biomolecular processes and the basic structure of the universe". We have replaced spacetime model with the time-invariant model. Seeing consciousness as something that appears in time is outdated. Linear phycological time "past-present-future" exists only in the human brain and consciousness is far beyond the brain and phycological time. The universe is time-invariant, time as duration enters existence when measured by the observer. In our model evolution of life has its information basis in the higher dimensional layers of SQS. Entire universe is existing in a time-invariant SQS, everything in the universe is entangled via time-invariant SQS. In $\mathbb{C}^4$SQS information transfer is of the light speed. In $\mathbb{C}^n$SQS information transfer is immediate. $\mathbb{C}^n$SQS is the medium of EPR-type entanglement. This is because the frequency of the photon in $\mathbb{C}^n$SQS that represent consciousness tends to infinite value and its wavelength tends to zero. Its velocity tends to zero ($v_{\to\infty}, \lambda_{\to 0}, v = v$ $\lambda = \to 0$). Consciousness is the carrier of the immediate information transfer. Back in 2014, Max Tegmark published an article where he discussed that consciousness could be understood as a state of mater. In the $\mathbb{C}^n$SQS model, all that exists in the universe is energy. Matter, electromagnetic energy, and consciousness are different aspects of the same energy. There is no need to think that matter is primary and consciousness is a state of matter or that consciousness is primary and the matter is its manifestation. They are both coexistent forms of the same energy. In $\mathbb{C}^n$SQS model dichotomy matter/consciousness is solved. Energies of all layers of $\mathbb{C}^n$SQS are interwoven. Seeing them separate seems wrong, they are one organism we name it "Universe".

Einstein and Bohm were not in the favour of the idea that the universe is a random phenomenon with no order. Einstein has proposed "hidden variables" to explain the EPR-type experiments, Bohm has proposed "implicate order of the universe", a model that proposes the universe is an intelligent system. There is a deep ontological similarity between Einstein's, Bohm's, and our view.

In our model universe is governed by consciousness and we humans have to search consciousness experientially in order to be able to follow cosmic laws and build human society accordingly. In our view today's quantum physics has limitations to describe consciousness because it sees consciousness as a phenomenon that is the domain of the "real world". Consciousness is a subjective phenomenon and as such "complex world". Adams and Petruccione are also pointing out the question of quantifying consciousness in the domain of quantum physics: "A formal description of consciousness, given the difficulty of quantifying its subjective experience, would likely borrow from complex network theory as well as disciplines from physics and philosophy. The question is still open as to whether quantum physics has something to add to the debate". In chapter 2 we developed the mathematical model of consciousness; however, we think that modelling of consciousness is not giving us the final answer about its nature. Consciousness can be known only by direct experience that is beyond rational modelling. Consciousness itself is of a higher cognitive phenomenon than thinking about it.

Self-organization is today recognized as a valid principle in developmental biology. It is well recognized that life is organizing itself. The mistake is to believe that this principle is ruling the development of life. No principle can rule a given process. A given principle in order to be real needs discovery of its physical origin. The principle of self-organization needs experimental verification. Our experiment "life-dead weight difference" proves that some higher dimensional type of $\mathbb{C}^n$SQS energies is present in the living organism. These higher-dimensional energies of $\mathbb{C}^n$SQS are the physical origin of self-organization. It makes no sense to see living organisms as an isolated system. Life is deeply related to the $\mathbb{C}^n$SQS.

Organic molecules have been found in the interstellar medium. In our model interstellar medium is the $\mathbb{C}^n$SQS. Molecules in interstellar areas have a tendency of self-organization because information of life is encoded in higher-dimensional layers of SQS. On the planets that are similar to the planet Earth, life has developed in intelligent beings. In our universe, there are many planes similar to our planet Earth where life could develop.

Bio-cosmology is a new branch of science that reaches beyond anthropocentrism and beyond geocentrism. We humans are not the centre of the universe. The evolution of life on Earth is the consistent part of a universal process that runs throughout the entire universe.

Particle physics needs rigorous re-examination

The fundamental question of particle physics is how elementary particles that have extremely short lifetimes can be constitutive "elements" of the proton. Applying system theory, we will take proton as a stable system that is composed out of elements that are elementary particles. How quarks and gluons that have extremely short lifetimes can build a proton is an important question. The hydrogen atom, for example, is a system composed out of two stable elements: proton and electron. How a proton could be a system made out of unstable elements as quarks and gluons is an unanswered question that particle physics need to face to strengthen its theoretical basis.

Proton and electron are stable systems. Their lifetime is close to the infinite. When it is not composed in the nucleus, Neutron is unstable; in about 15 minutes, the neutron will fall apart into proton and electron and antineutrino. From the view of system theory neutron is not the element of the system we call "atom". Atoms are composed only out of protons and electrons which are their real constitutive elements. Neutron is not a real particle because it is not stable. Stability in time is the inherent property of all elementary particles that build up an atom. An element of the atom that has a lifetime of 15 minutes cannot be thought of as its constitutive building element.

	orbit	nucleous
H	1 electron	1 proton
He	2 electrons	4 protons + 2 electrons
Li	3 electrons	6 protons + 3 electrons
Be	4 electrons	8 protons + 4 electrons
B	5 electrons	10 protons + 5 electrons
	N electrons	$2N$ protons + N electrons

Inside the nucleus of the Helium, we have 2 protons and 2 neutrons, which are composed out of 2 protons and 2 electrons. In system theory, Helium is composed of 8 stable elements: 4 protons and 4 electrons. Neutrons are not considered being elements of the Helium nucleus because they are not stable.

In the principle, elementary particles are building a stable system we call "proton" that is made out of several elements: two upper quarks, one down quark, and the unknown number of gluons which are supposed to glue together quarks.

Quarks lifetime is about 10^{-23} second. About gluons lifetime there is no data in the literature. They are not supposed to be stable particles. Miklos Gyulassy has reported in his article published back in 1996 in Nuclear Physics A, that the lifetime of quark-gluon plasma has a value about of zero.

In the "system theory" is known that a given system is stable if it is composed out of stable elements. The system theory regular question is: "How a proton which is a stable system may be a system made out of unstable elements"? Neutron is not a system because it is not stable despite it is composed out of stable elements. From the view of system theory, we need to understand the physical origin of proton stability. Particle physics needs to discover what mechanism makes proton stable. This is a challenge for particle physics from the perspective of system theory.

The hydrogen atom is a system with two stable components: the proton is positively charged, the electron is negatively charged, electromagnetic force holds them together. Proton is made out of many unstable elements. How unstable gluons are gluing together unstable quarks and make proton stable is important to understand. In technical terms, gluons are vector gauge bosons that mediate strong interactions of quarks in quantum chromodynamics (QCD). How the intermediation of unstable elements (gluons) between unstable elements (quarks) makes proton stable is an open question. From the view of system theory, QCD is a theoretical description that is not giving a satisfying answer about the physical origin of proton stability.

From the view of a system theory a given system to exist, to be stable, must be composed out of stable elements. For example, the table to be stable must have all 4 legs stable. One leg unstable means the table is unstable. In a group of people where only one person is unstable, the group can become unstable. You imagine you take away Saturn out of our Solar system. The system would become unstable. The proton is a special case of a system where all elements are unstable. What gives proton stability, particle physics does not know yet. If this missing mechanism that makes the proton stable will not be found, we can assume that quarks and gluons are only the epiphenomena existing in accelerated protons collisions and do not exist in the proton at rest. My proposal is, that the proton is a vortex of superfluid quantum space and has no constitutive elements. The model of a proton as a vortex of superfluid quantum space is proposed by the Russian physicist Valeriy I. Sbitnev back in 2017.

Proton as a vortex of superfluid quantum space

In my view, the idea of the proton as an isolated element in the space seems false. Proton and electron are both vortexes of the superfluid quantum space. Erving Schrodinger also regarded quantum space as the fundamental energy of the universe. He used to say: "What we observe as material bodies and forces are nothing but shapes and variations in the structure of space".

On the table below we have discovered quarks and bosons and their lifetimes. They all have an extremely short lifetime.

Top quark	171,2 GeV/c2	10E-23 s	1995
Higgs boson	125 GeV/c2	1,56 x 10E-22 s	2013
Z boson	91 GeV/c2	3 x 10E-25 s	1983
W boson	80 GeV/c2	3 x 10E-25 s	1983
Bottom quark	4,2 GeV/c2	10E-23 s	1977
Charm quark	1,27 GeV/c2	10E-23 s	1974
proton	938 MeV/c2	stable lifetime	1886
strange quark	104 MeV/c2	10E-23 s	1968
Down quark	4,8 MeV/c2	10E-23s	1968
Up quark	2,4 MeV/c2	10E-23 s	1868

How these unstable particles could build proton for me personally is a mystery that is not clear yet. In my opinion, most of the elementary particles "discovered" in cyclotrons are the momentary fluxes of energy that are released by particle collisions.

For example, when we accelerate proton close to the light speed its relativistic energy increases about 7460 times. The proton rest mass is $m_0 = 1.672 \cdot 10^{-27} kg$. In an accelerator, the proton relativistic energy reaches in terms of rest mass m_0 a value of $E = m_0 \cdot c^2 \cdot 7460$. When accelerated, a proton is absorbing in its vortex the energy of superfluid quantum space. When a collision is happening this absorbed energy is released as the many different "sparkles" that immediately turn back into the superfluid quantum space. Most of the particles discovered in cyclotrons are these sparkles.

Vortex model of elementary particles and the double-slit experiment

Let's imagine a double-slit experiment in the model of vortex model where particles are moving only through the upper slit. We are sending electrons through the upper slit. The interference pattern will appear also behind the down slit.

Physics is searching for 100 years to solve this puzzle. It can be solved only with the understanding that space is not "empty", space is the fundamental energy of the universe, in physics, we call it "superfluid quantum space" Moving particles which are vortexes of space are creating waves of space and these waves are creating interference patterns behind the down slit. A given particle in motion will always create a wave in the superfluid quantum space.

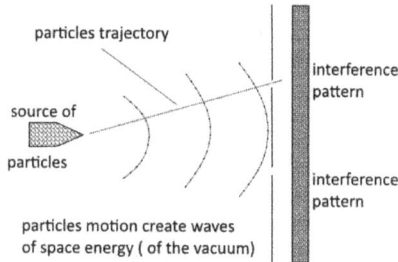

In general double-slit experiment has nothing to do with the superposition and with the observer. It is a pure technicality of the physical world.

The existence of antimatter in physical reality is questionable

Cyclotron physics in general is not falsifiable and is not bijective. It will never give any technological application. Positron and antiproton are two discovered particles accordingly to the supersymmetry model. Positron is the antiparticle of the electron. It has been found in cosmic rays, but in the contact with ordinary matter is unstable. It has a lifetime of about 10^{-10} seconds. Antiprotons are also found in cosmic rays. They are unstable, they are typically short-lived since any collision with a proton will cause both particles to be annihilated in a burst of energy. Their lifetime, when kept in artificially made physical circumstances in the lab is about 32 hours. In system theory, unstable elements cannot build a stable system. The idea that antimatter is existing in the physical universe is unproven speculation. CERN research on antihydrogen will not give positive results because the idea of antimatter existence is based on wrong assumptions. We can read on the CERN home page: "In 1928, British physicist Paul Dirac wrote down an equation that combined quantum theory and special relativity to describe the behaviour of an electron moving at a relativistic

78

speed. The equation – which won Dirac the Nobel Prize in 1933 – posed a problem: just as the equation $x^2 = 4$ can have two possible solutions (x = 2 or x = −2), so Dirac's equation could have two solutions, one for an electron with positive energy, and one for an electron with negative energy. But classical physics (and common sense) dictated that the energy of a particle must always be a positive number. Dirac interpreted the equation to mean that for every particle there exists a corresponding antiparticle, exactly matching the particle but with opposite charge. For example, for the electron there should be an "antielectron", or "positron", identical in every way but with a positive electric charge. The insight opened the possibility of entire galaxies and universes made of antimatter".

Yes, the equation $x^2 = 4$ has two possible solutions. But we cannot use this mathematical fact in cosmology and think that galaxies made out of antimatter exist. Between this equation and the physical world, there is no bijectivity, the entire idea of antimatter is not falsifiable. Particle physics needs re-examination to fulfill the requests of system theory which requires bijectivity as the necessary standard for a given theory to become an adequate model of the physical world.

The Higgs mechanism is an unnecessary complication

The Higgs mechanism was introduced to give some of the elementary particles, like proton, inertial mass m_i. Higgs field interacts with some particles and "slow" them down giving them inertial mass and does not interact with some other particles for example with the photon that has no inertial mass. The original idea of theoretical physics was to establish the model that will explain the inertial mass of some elementary particles.

In today's mainstream physics inertial mass m_i is equal to rest mass m_0. They are thought to be one single phenomenon. Here the problem has been raised. Because today Higgs mechanism is understood as a field that is giving the rest mass of the elementary particles.

Higgs field interacts with the quarks but does not interact with the gluons that are 99% of the proton rest mass. Quarks are 1% of the proton rest mass. How Higgs field could give the rest mass to the proton remains an unanswered question.

The Higgs mechanism was introduced to give some of the elementary particles, as for example, proton, inertial mass m_i. Higgs field interacts with some particles and "slow" them down giving them inertial mass and does not interact with some other particles as for example with the photon that has no inertial mass. The original idea of theoretical physics was to establish the model that will explain the inertial mass of some elementary particles.

In today's mainstream physics inertial mass m_i is equal to rest mass m_0. They are thought to be one single phenomenon. Here the problem has been raised. Because of this misunderstanding Higgs mechanism is today understood as a field that is giving the rest mass of the elementary particles.

Higgs field interacts with the quarks but does not interact with the gluons that are 99% of the proton rest mass. Quarks are 1% of the proton rest mass. How Higgs field could give the rest mass to the proton remains an unanswered question.

We have shown in chapter 2 that inertial mass is the result of the rest mass interaction with the SQS. Rest mass m_0 is diminishing the energy density of SQS in its centre exactly for the amount of the rest mass. The outer higher pressure of SQS is pushing towards the centre of the rest mass where the energy density of SQS is minimal. The pressure of SQS is inertial mass m_i.

With the introduction of the Higgs field, we have now three basic fields in physics: Higgs field, gravity field, and electromagnetic field. And we do not know exactly how are related. And we will never know. Because there is only one field of SQS. Its variable energy density carries gravity and inertia, its excitation carries electromagnetism. As Albert Einstein told us: "Everything should be made as simple as possible, but not simpler".

Author wishes to thank the main research fellows and co-authors Štefan Čelan, Davide Fiscaletti, Saeid Jafari, and Aram Bahroz Brzo for their contribution to our research project "Advances of Relativity and Cosmology".

References

Šorli, A., & Čelan, Štefan. (2020). The End of Space-time: Physics-Mathematics. *International Journal of Fundamental Physical Sciences*, *10*(4), 31-34. https://doi.org/10.14331/ijfps.2020.330139

Šorli , A. S. ., & Čelan , Štefan. (2021). Einstein's Misunderstanding of Time in the Time-Invariant Universe: Astrophysics, Relativity. *International Journal of Fundamental Physical Sciences*, *11*(1), 1-5. https://doi.org/10.14331/ijfps.2021.330143

Šorli, A.S. & Čelan Š. Superfluid quantum space as the unified field theory, Reports in Advances of Physical Sciences, Vol. 4, No. 3 (2020) 2050007, https://doi.org/10.1142/S2424942420500073 (2021)

Šorli A., Čelan Š., Advances of Relativity Theory, Physics Essays, Vol 34: Pages 201-210 (2021) http://dx.doi.org/10.4006/0836-1398-34.2.201

Amrit Srecko Šorli, Štefan Čelan, Saeid Jafari, Davide Fiscaletti, & Aram Bahroz Brzo. (2021). Multiverse in Dynamic Equilibrium. http://doi.org/10.5281/zenodo.4741540

Amrit Srecko Šorli, & Štefan Čelan. (2021). Time as the result of observer's measurement. http://doi.org/10.5281/zenodo.4761107

Sorli, A.S.; Čelan, Š. Biocosmology - Multiverse, Life, and Consciousness. *Preprints* 2019, 2019120402

Sorli, A. S. (2020). System Theory, Proton Stability, Double-Slit Experiment, and Cyclotron Physics. *JOURNAL OF ADVANCES IN PHYSICS*, *17*, 161–168. https://doi.org/10.24297/jap.v17i.866

Amrit S. Sorli, Stefan Celan Schwarzschild energy density of superfluid quantum space and mechanism of AGNs' jets Advanced Studies in Theoretical Physics, Vol. 15, 2021, no. 1, 9-17 doi: 10.12988/astp.2021.91506

www.ingramcontent.com/pod-product-compliance
Lightning Source LLC
Chambersburg PA
CBHW061838220326
41599CB00027B/5325